New Technologies in Health Care

Health, Technology and Society

Series Editors: Andrew Webster, University of York, UK and Sally Wyatt, University of Amsterdam, The Netherlands

Titles include:

Andrew Webster (*editor*)
NEW TECHNOLOGIES IN HEALTH CARE
Challenge, Change and Innovation

Forthcoming titles include:

Gerard de Vries and Klasien Horstman
GENETICS FROM LABORATORY TO SOCIETY

Herbert Gottweis, Brian Salter and Catherine Waldby
THE GLOBAL POLITICS OF THE HUMAN EMBRYONIC STEM CELL SCIENCE

Steven P. Wainwright and Clare Williams
THE BODY, BIO-MEDICINE AND SOCIETY
Reflections on High-Tech Medicine

Alex Faulkner
MEDICAL DEVICES AND HEALTHCARE INNOVATION

Andrew Webster
HEALTH, TECHNOLOGY AND SOCIETY
A Sociological Critique

Health, Technology and Society
Series Standing Order ISBN 1–4039–9131–6 Hardback
(*outside North America only*)

You can receive future titles in this series as they are published by placing a standing order. Please contact your bookseller or, in case or difficulty, write to us at the address below with your name and address, the title of the series and the ISBN quoted above.

Customer Services Department, Macmillan Distribution Ltd, Houndmills, Basingstoke, Hampshire RG21 6XS, England

---

New Technologies in Health Care

Challenge, Change and Innovation

Edited by

Andrew Webster
University of York

Selection and editorial matter © Andrew Webster 2006
Chapters © the authors 2006

All rights reserved. No reproduction, copy or transmission of this publication may be made without written permission.

No paragraph of this publication may be reproduced, copied or transmitted save with written permission or in accordance with the provisions of the Copyright, Designs and Patents Act 1988, or under the terms of any licence permitting limited copying issued by the Copyright Licensing Agency, 90 Tottenham Court Road, London W1T 4LP.

Any person who does any unauthorised act in relation to this publication may be liable to criminal prosecution and civil claims for damages.

The authors have asserted their rights to be identified as the authors of this work in accordance with the Copyright, Designs and Patents Act 1988.

First published in 2006 by
PALGRAVE MACMILLAN
Houndmills, Basingstoke, Hampshire RG21 6XS and
175 Fifth Avenue, New York, N.Y. 10010
Companies and representatives throughout the world.

PALGRAVE MACMILLAN is the global academic imprint of the Palgrave Macmillan division of St. Martin's Press, LLC and of Palgrave Macmillan Ltd. Macmillan® is a registered trademark in the United States, United Kingdom and other countries. Palgrave is a registered trademark in the European Union and other countries.

ISBN-13: 978–1–4039–9130–0 hardback
ISBN-10: 1–4039–9130–8 hardback

This book is printed on paper suitable for recycling and made from fully managed and sustained forest sources.

A catalogue record for this book is available from the British Library.

Library of Congress Cataloging-in-Publication Data
    New technologies in health care : challenge, change, and innovation / editor, Andrew Webster.
      p. cm.—(Health, technology, and society)
    Includes bibliographical references and index.
    ISBN 1–4039–9130–8 (cloth)
     1. Medical instruments and apparatus. 2. Technology assessment.
I. Webster, Andrew. II. Series.

R855.3.N49 2006
610.28—dc22
                                                             2005056486

10 9 8 7 6 5 4 3 2 1
15 14 13 12 11 10 09 08 07 06

Printed and bound in Great Britain by
Antony Rowe Ltd, Chippenham and Eastbourne

Contents

List of Tables	vii
List of Figures	viii
Acknowledgements	ix
Series Editors' Preface	x
List of Contributors	xi

Introduction: New Technologies in Health Care: Opening the Black Bag
Andrew Webster — 1

Part 1   Genetic Risk, Reproduction and Identity

1 The Genetic Iceberg: Risk and Uncertainty — 11
   Aditya Bharadwaj, Lindsay Prior, Paul Atkinson, Angus Clarke and Mark Worwood

2 Navigating the Troubled Waters of Prenatal Testing Decisions — 25
   Gillian Lewando Hundt, Josephine Green, Jane Sandall, Janet Hirst, Shenaz Ahmed and Jenny Hewison

3 Genetic Ambivalence: Expertise, Uncertainty and Communication in the Context of New Genetic Technologies — 40
   Anne Kerr and Sarah Franklin

Part 2   Information and Empowerment

4 'Pathways to the Doctor' in the Information Age: the Role of ICTs in Contemporary Lay Referral Systems — 57
   Sarah Nettleton and Gerard Hanlon

5 Desperately Seeking Certainty: Bone Densitometry, the Internet and Health Care Contexts — 71
   Eileen Green, Frances Griffiths, Flis Henwood and Sally Wyatt

6 Telemedicine, Telecare and the Future Patient: Innovation, Risk and Governance — 84
   Tracy Finch, Carl May, Maggie Mort and Frances Mair

7	Patient 'Expertise' and Innovative Health Technologies *Katie Ward, Mark Davis and Paul Flowers*	97
8	Making Sense of Mediated Information: Empowerment and Dependency *Joe Cullen and Simon Cohn*	112

Part 3 Innovation, Context and Meaning

9	Time, Place and Settings: Negotiating Birth, Childhood and Death *Jane Seymour, Elizabeth Ettorre, Janet Heaton, Gloria Lankshear, David Mason and Jane Noyes*	131
10	Replacing Hips and Lenses: Surgery, Industry and Innovation in Post-War Britain *J. S. Metcalfe and John Pickstone*	146
11	Access, Agency and Normality: the Wheelchair and the Internet as Mediators of Disability *Susie Parr, Nick Watson and Brian Woods*	161

Part 4 Regulation and Evaluation of IHTs

12	Understanding the 'Productivity Crisis' in the Pharmaceutical Industry: Over-regulation or Lack of Innovation? *Paul Martin, John Abraham, Courtney Davis and Alison Kraft*	177
13	Regulating Hybridity: Policing Pollution in Tissue Engineering and Transpecies Transplantation *Nik Brown, Alex Faulkner, Julie Kent and Mike Michael*	194
14	Cultural Politics and Human Embryonic Stem Cell Science *Brian Salter*	211
15	Regulation and the Positioning of Complementary and Alternative Medicine *John Chatwin and Philip Tovey*	224
16	Evaluation as an Innovative Health Technology *David Armstrong*	232

Bibliography	242
Author Index	268
Subject Index	274

List of Tables

1.1	Sample of 'patients' interviewed in the Cancer Genetics Study	14
2.1	Genetic screening: Information provided to women	28
2.2	Reported sources of information about Down's syndrome	28
2.3	Perceptions of Down's syndrome	29
6.1	Telemedicine and telehealthcare: Innovation, risk and governance	94
12.1	Productivity in the pharmaceutical industry 1951–1970	188
14.1	Regulations in EU Member States regarding human embryonic stem cell research (March 2003)	216

List of Figures

4.1	The future of medical education: Speculation and possible implications	59
4.2	Consultation as a social process	64
10.1	The eye as a design space	148
10.2	The hip joint and artificial implants	151
12.1	Licensing applications for new active substances in the EU under the centralised procedure and the mutual recognition procedure – 1998 to 2003	178
12.2	10-year trends in major drug and biological product submissions to the FDA	179
12.3	Number of NMEs first launched onto the world market	180
12.4	Growth of pharmaceutical R&D	189
14.1	Bioethics and the units of cultural trading	217
14.2	Major ethical components of the political narrative	218

Acknowledgements

Health technologies may claim to be radically new and distinctive. In this book, though, any novelty they have has been brought to life through long-established collaborative efforts among the members of the thirty-one research teams who have co-written the chapters in this book. The authors' willingness to identify themes and issues that cut across their respective projects has made my task as editor that much easier and more rewarding. The book captures some of the key results of a five-year research programme, but not all: the full details are available on the Innovative Health Technologies programme website (www.york.ac.uk/res/iht).

Over the past five years, apart from the research teams themselves, there are many who have made my job as Programme Director a real pleasure, and when things got more tricky, helped to resolve matters with professional skill and good sense. Within the ESRC itself, the principal source of funding for the Programme, I want to thank Ian Diamond, Ros Rouse, Joy Todd and Board member Waqar Ahmad who have been extremely supportive throughout the five years. I am also extremely grateful to all the members of the Programme Advisory Committee: Dame Mary Warnock, Clara Mackay (consecutive Chairs), Ray Fitzpatrick, Peter Glasner, Peter Greenaway, Gareth Lloyd Jones, Clare Matterson, Nikki Ratcliffe and George Sarna. Peter Dukes was especially helpful in early discussions with the MRC, the other primary source of funding, while Malcolm Skingle, from GSK, has also provided invaluable support.

At York, I want to express my gratitude to all those in my research unit, SATSU, for their support. I want especially to thank my two colleagues Stephanie Hazel-Gant, Programme Administrator, and Luana Pritchard, Programme Secretary, who have played a key role in keeping the programme on track and me relatively sane. I thank them not only for their extremely effective administration but also for carrying out their responsibilities with a ready helpfulness, combining professional skill and friendship with ease and great charm.

I want to thank Jill Lake at Palgrave Macmillan who, as publisher of this book and the forthcoming series texts, has been a pleasure to work with, and to Christine Ranft for her skills as copy-editor.

Finally, I would like to thank my family, especially Helen, who has given me more support and understanding than should be asked of anyone, while Matt and Nick have helped keep both our minds on things other than work – a distraction more welcoming than any health technology.

Andrew Webster
University of York
December 2005

Series Editors' Preface

Medicine, health care, and the wider social meaning and management of health are undergoing major changes. In part this reflects developments in science and technology, which enable new forms of diagnosis, treatment and the delivery of health care. It also reflects changes in the locus of care and the burden of responsibility for health. Today, genetics, informatics, imaging and integrative technologies, such as nanotechnology, are redefining our understanding of the body, health and disease; at the same time, health is no longer simply the domain of conventional medicine, nor the clinic.

More broadly, the social management of health itself is losing its anchorage in collective social relations and shared knowledge and practice, whether at the level of the local community or through state-funded socialised medicine. This individualisation of health is both culturally driven and state sponsored, as the promotion of 'self-care' demonstrates. The very technologies that redefine health are also the means through which this individualisation can occur – through 'e-health', diagnostic tests, and the commodification of restorative tissue, such as stem cells, cloned embryos and so on.

This Series explores these processes *within* and *beyond* the conventional domain of 'the clinic', and asks whether they amount to a qualitative shift in the social ordering and value of medicine and health. Locating technical developments in wider socio-economic and political processes, each text discusses and critiques recent developments within health technologies in specific areas, drawing on a range of analyses provided by the social sciences. Some will have a more theoretical, others a more applied focus, interrogating and contributing towards a health policy. All will draw on recent research conducted by the author(s).

The Health, Technology and Society Series also looks toward the medium term in anticipating the likely configurations of health in advanced industrial societies and does so comparatively, through exploring the globalisation and the internationalisation of health, health inequalities and their expression through existing and new social divisions.

Andrew Webster and Sally Wyatt

List of Contributors

John Abraham is Professor of Sociology and Co-director of the Centre for Research in Health and Medicine (CRHaM) at the University of Sussex

Shenaz Ahmed is a Research Fellow in the Academic Unit of Psychiatry and Behavioural Sciences at the University of Leeds

David Armstrong is Reader in Sociology as Applied to Medicine in the Department of General Practice and Primary Care at King's College School of Medicine, London

Paul Atkinson is Professor of Sociology at the Cardiff School of Social Sciences, Cardiff University

Aditya Bharadwaj is Lecturer in Sociology in the School of Social and Political Studies, University of Edinburgh

Nik Brown is Lecturer in Sociology and Deputy Director of the Science and Technology Studies Unit (SATSU) in the Department of Sociology at the University of York, UK

John Chatwin is a Research Fellow at the School of Healthcare, University of Leeds

Angus Clarke is Professor in Clinical Genetics, Cardiff University, and directs the Cardiff MSc course in genetic counselling

Simon Cohn is a Lecturer in Anthropology, Goldsmiths College, University of London

Joe Cullen is Principal Researcher and Consultant at The Tavistock Institute, London

Courtney Davis is a Research Fellow in Sociology and at the Centre for Research in Health and Medicine at the University of Sussex

Mark Davis is Senior Lecturer in the School of Social Science, Media and Cultural Studies, UEL, London

Elizabeth Ettorre is Professor of Sociology and Associate Dean of Research and Enterprise, Faculty of Social Sciences and Business, University of Plymouth

Alex Faulkner is Senior Research Associate at the Cardiff Institute of Society, Health and Ethics in the School of Social Sciences, Cardiff University

Tracy Finch is Senior Research Fellow in the Health Technologies and Human Relations research group at the Centre for Health Services Research (CHSR), University of Newcastle

Paul Flowers is Reader in the Department of Psychology, School of Life Sciences, Glasgow Caledonian University

Sarah Franklin is Professor in Sociology at BIOS, London School of Economics

Eileen Green is Professor of Sociology and Director of the Centre for Social and Policy Research in the School of Social Sciences and Law at the University of Teesside

Josephine Green is Professor of Psychosocial Reproductive Health and Deputy Director of the Mother & Infant Research Unit at the University of York

Frances Griffiths is Deputy Director of the Centre for Primary Health Care and Clinical Senior Lecturer at the University of Warwick

Gerard Hanlon is Professor of Organisation and Society at the Management Centre, University of Leicester

Janet Heaton is Research Fellow at the Social Policy Research Unit, University of York

Flis Henwood is Reader in Social Informatics in the School of Computing, Mathematical and Information Sciences, University of Brighton

Jenny Hewison is Professor of the Psychology of Health Care, Academic Unit of Psychiatry and Behavioural Sciences, University of Leeds

Janet Hirst is a Research Fellow in the Mother and Infant Research Unit, Department of Health Sciences at the University of Leeds

Julie Kent is Reader in Sociology in the School of Sociology, University of the West of England, Bristol

Anne Kerr is Professor in the Department of Sociology and Social Policy, University of Leeds

Alison Kraft is Senior Research Fellow of IGBIS, Department of Sociology and Social Policy, University of Nottingham, UK

Gloria Lankshear is Postdoctoral Research Fellow in the Department of Health and Social Work, University of Plymouth

Gillian Lewando Hundt is Professor of Social Sciences in Health and Co-director of the Institute of Health, School of Health and Social Studies, University of Warwick

Frances Mair is Professor of Primary Care Research in the Division of Community Health Sciences, University of Glasgow Medical School, UK

David Mason is Professor of Sociology and Dean of the School of Social Sciences, Nottingham Trent University

Paul Martin is Deputy Director of IGBIS, Department of Sociology and Social Policy, University of Nottingham

Carl May is Professor of Medical Sociology and leads the Health Technologies and Human Relations research group at the Centre for Health Services Research, University of Newcastle, UK

Stanley Metcalfe is the Stanley Jevons Professor of Political Economy and Executive Director of the Centre for Research on Innovation and Competition, University of Manchester

Mike Michael is Professor of Sociology of Science and Technology at the Department of Sociology, Goldsmiths College, University of London

Maggie Mort is Senior Lecturer in the Institute for Health Research and Co-director of the Centre for Science Studies, University of Lancaster, UK

Sarah Nettleton is a Senior Lecturer in Social Policy, in the Department of Social Policy and Social Work at the University of York

Jane Noyes is a Lecturer at the School of Nursing, Midwifery and Social Work, University of Manchester

Susie Parr is Research and Development Co-ordinator for Connect, the UK communication disability network

John Pickstone is Professor and Director of the CHSTM/Wellcome Unit for the History of Medicine, University of Manchester

Lindsay Prior is Professor of Sociology at the Queen's University Belfast, Belfast

Brian Salter is Professor of Biopolitics in the Institute of Health at the University of East Anglia

Jane Sandall is Professor of Midwifery and Women's Health at King's College, London and co-leads the Women and Family Health Research Group

Jane Seymour is Sue Ryder Professor of Palliative and End of Life Studies at the Sue Ryder Centre, School of Nursing, University of Nottingham

Philip Tovey is Reader in Health Sociology, School of Healthcare, University of Leeds

Katie Ward is a Research Officer at ScHARR, University of Sheffield

Nick Watson is Professor of Disability Studies and Director of the Strathclyde Centre for Disability Research, at the University of Glasgow

Andrew Webster (IHT Programme Director) is Professor in the Sociology of Science and Technology and Director of the Science and Technology Studies Unit, Department of Sociology, University of York

Brian Woods is Research Fellow at the Science and Technology Studies Unit, Department of Sociology at the University of York

Mark Worwood is Professor in Haematology at the University of Wales College of Medicine, Cardiff

Sally Wyatt is based in the Amsterdam School of Communications Research (ASCoR) at the University of Amsterdam

Introduction: New Technologies in Health Care: Opening the Black Bag

Andrew Webster

Reading the first page of a new book, as you are now, is a bit like being confronted by a new medical technology: you're not sure whether it will be good for you, how unpleasant (or pleasant) the experience may be, and where you might end up. This book should make you worry, but hopefully, not ill. On the other hand, a book is pretty familiar territory: you expect it to be similar to many other books you've read in terms of its structure and general appearance, and in the end it should have some benefit for you. An innovative health technology (IHT) is a similar mix of the strange, uncertain and perhaps forbidding, combined with a sense that you've seen or experienced something like it before. A diagnostic (say genetic) test, a hip replacement, an ultrasound scan, a mammogram and so on, have been available for decades yet today's versions are qualitatively distinct from and much more extensive in range than those of the past. Perhaps most importantly, their technical sophistication generates both greater accuracy and use but also reveals new pathological uncertainties and creates new personal and professional risks. It's not clear what the ending of the book will turn out to be, and even when you get there you can be left hanging.

New medical technologies and techniques are being developed at a rapid pace by clinicians, surgeons, bio-scientists, medical researchers, medical device technologists, pharmaceutical companies and so on. They are also being found more widely distributed in both clinical settings – as instruments, diagnostic tools and techniques are acquired and diffuse through hospitals and doctors' surgeries – and non-clinical ones too. The latter includes the home as more and more 'near-patient' devices are used, or technologies once found only in hospitals are reconfigured for home use, the gym, and even the shopping mall where, as in the US, step-in booths provide whole-body readouts on the apparent state of our health.

The diversity and ubiquity of forms of medical technique, device or product can be seen to be part of the process of *medicalisation* that many sociologists of health and medicine have described (see, for example, Foucault, 1975; Waldby, 1997; Turner, 1996) and what Pickstone (2002b: 3) has called the advent of '*consumerist* medicine', whereby medicine becomes 'a commodity, chosen by individuals, usually in free markets'. While medicine was once the primary preserve of welfare and state provision, both processes mean a broadened (indeed globalised) as well as more specialised, segmented market for medicine. At the same time, the ubiquity of medicine and its forms enables, as Lupton says, 'the penetration of the clinical gaze into the everyday lives of citizens' (Lupton, 1997: 107). As a result, 'surveillance medicine' (Armstrong, 2002) acts to monitor, discipline and organise both the individual and social body – whether in the form of the electronic patient record, personal DNA biobanks, the regulation of public health, sexuality, diet, or a subscription to healthier lifestyles. Technologies of screening and testing are key means through which such surveillance is enabled and secured.

At the same time, the broadening and opening up of medicine and health technologies beyond the clinic – and the doctor's 'black bag' – has meant that the patient and consumer of health care can access and to some degree construct their own definitions of what it means to be healthy, or to construct personal narratives of health built, for example, via the web or personally customised health products, services and devices. Medicine and its meaning becomes less unified and increasingly diverse (see Berg and Mol, 1998).

In broader terms, we can say that the status of health and of illness is being redefined today whether from within medicine itself or through the actions and narratives of the health consumer (Webster, 2002). As medicine and its increasingly sophisticated technology penetrates to the depths of pathologies, it produces, somewhat paradoxically, increasingly marginal judgements of the severity or otherwise of disease – as in the diagnosis of cancer, or the likely onset of multifactorial genetic illnesses. The treatment of 'risk' is much more problematic than the treatment of a well-specified disorder. As health consumers engage with prospective health risks as much as everyday aches and pains, 'being well' takes on different dimensions, timeframes and layers of meaning.

To the extent that this is the case, notions of normality and abnormality become less well-defined – the language of 'risk profiles' rather than simply 'symptoms' is used – and thereby precisely what *is* being medicalised and what subject to medical surveillance less easily characterised. Medical diagnosis, ever a site for contest and debate, is also a site through which medical resources are deployed; but as risk calculation becomes the norm, evaluating the merits of using medical resources in one way rather than another becomes a difficult question for health providers, whether at the level of the clinic or the state. The response of state agencies worldwide has been to try to tighten up procedures through which health provision is evaluated, health technologies assessed or new programmes introduced, while at the same time passing much greater responsibility to health consumers to determine and pay for their health needs.

These developments are not entirely the result of innovative health technologies. Expertise, whether medical or otherwise, is more subject to lay challenge in late modernity (Giddens, 1990) such that alternative accounts of health and illness proliferate. Relative affluence in advanced industrial states produces new forms of chronic disorder, such as obesity, which occupy an uncertain position on a threshold that marks medical intervention off from lifestyle 'choices'. Finally, the ageing population creates major demands for health care systems with respect to the management of chronic illnesses associated with the latter stages of the life course. This has led government to search for new ways to manage ageing outside of clinical settings (especially through a combination of telehealthcare and the placing of more responsibility on informal carers).

These wider cultural and structural shifts are of central importance in understanding the context within which innovative health technologies are deployed today. Even so, there are some specific features of these technologies that accelerate the processes through which the meaning of health and illness is redefined and made more provisional, thereby changing a person's relationship to their own body. Four features seem to be particularly characteristic of new health technologies.

First, there is a growing range of technology that is designed to be *embedded* within the body, in the form of smart prosthetics that enable the monitoring and thereby the clinical management of illness. However, while this may improve the degree of control over a specific condition, the specificity of the condition can become blurred since real-time 24/7 monitoring of, say a heart complaint, can mean that patients and doctors find it difficult to distinguish between an 'innocent condition' and one that is a cause for concern.

Secondly, contemporary technologies are *projective* of the body and its pathologies across time and space via complex socio-technical systems, such as, in the UK, the national information spine being built within the NHS. The reconfiguring of patient information and symptomatology as digital data enables the mobilisation and standardisation of patient information across different clinical sites. In doing so, the patient's status is crucially dependent on its representation via digital fields and their associated databases. At the same time, as is commonly experienced elsewhere, such systems can be corrupted or lose their integrity, become inoperable or poorly maintained. The digital version of the patient's health may bear little resemblance to how they actually feel.

Third, new health technologies are increasingly *hybridised* inasmuch as they draw on science and technology from across diverse fields anchored in very different experimental, instrumental and evaluative paradigms. Tissue engineering that draws on the wet biology of stem cells and the engineering of synthetic materials is a case in point, such that getting tissue engineering to 'work' (in the form, for example, of 'scaffolds' for bone replacement) depends on researchers and clinicians constructing hybrid definitions of what counts as an effective working scaffold and what is therefore to be deemed to be a clinical 'success'.

Finally, the power of digital systems to generate visual or pictorial images of the body and disease pathologies has meant that the *representational* power of new health technologies is of major importance in describing and agreeing on the status of a person's health or ill-health. Highly colourful computerised representations of genetic information, graphics of the virtual patient used for clinical training, telemedical images of skin lesions, ultrasound scans of the foetus – all provide new markers and clues for making sense of clinical disorders, but also do so in such a way that can create new sources of disagreement about the same. This may be linked to professional disputes over how to read such images, or to greater detail (as in DNA graphics) such that, paradoxically, the threshold of normal and abnormal becomes less easy to define.

These four features – *embedded, projective, hybridised* and *representational* – work to change the boundaries and meaning of health and disease. Moreover, each feature is not discrete unto itself, but links to the other three to a greater or lesser extent: for example, embedded technologies often depend on projective systems beyond the body to maximise their utility, which in turn may well involve new representational forms of the pathology they identify.

These characteristics of innovative health technologies are explored in this volume, which brings together the work of interdisciplinary research across the social and other sciences. While the substantive focus covers a broad canvas, there are some common themes that form the conceptual framework of the book. Most important of all is that technologies are examined as mediated by and an expression of *social relationships*. What is therefore 'innovative' about them cuts across the social, the material and the technical. There is in other words, no sense in which technologies carry intrinsic properties or have uniform effects in their application, or develop in some progressive, linear way. They do, however have material agency – a prosthetic device, a genetic screening, a new drug and so on, have effects, and are designed in very specific ways to produce those rather than any other. However, we cannot know decisively and precisely what such effects may be in advance, a point argued most forcefully by Pickering (1995). Instead, such agency is always emergent and depends on the ways in which, in practice, human and material agency are intertwined in what he calls the 'mangle of practice' (21). As he puts it elsewhere, 'neither we nor nature call all of the shots ... we live in a space of inexhaustible becoming' (2001: 5).

This provisionality also means that technologies are difficult to control and manage not only in an immediate technical sense but also socially and politically. The world of institutionalised science (and medicine) often make claims on science's behalf which prove to be unsustainable, and lay people are increasingly aware of the unpredictable consequences and risks of new technologies. Patients directly experience the unintended effects of scientific medicine – such as the adverse effects of drugs – but may also be enrolled by medical research in highly experimental trials whose results are by definition uncertain. These are, of course, two very different types of risk, the second more transparent than the first, but both illustrate in different ways Pickering's observation about who or what is 'calling the shots'. Ultimately, what 'shots' get called will depend on value judgements, economic considerations and technical possibilities – all contested and contestible. The greater the technological possibilities opened up by innovation the more important it becomes that we ask not only can, but *why* should one road be taken rather than another (see Mol, 2002; Hoffmann, 2001).

Organisation of the book

The volume is organised in four Parts according to a number of core research themes that figured prominently in the IHT research programme. Most of the chapters bring together researchers from more than one project (in one case three), and have been prepared in such a way as to draw out common or contrasting issues from the very different empirical work that the authors undertook. This means that the chapters do not reflect the full range of findings that each project generated, but rather illustrate how we can read across and integrate results at a more thematic level. The material provides the first major collection of work that forms the foundation for a new social science of health technologies.

Part 1, *Genetic Risk, Reproduction and Identity*, carries three chapters that in their different ways explore the uncertainties, ambiguities and risks associated with contemporary genetics in clinical and non-clinical contexts. The chapter by Bharadwaj et al. explores the meaning of genetic risk and describes what the authors call a genetic iceberg, 'a pyramid of potential pathology' of those who may well be identified as carriers of genetic risk without any illness symptoms. They show how the threshold, the boundary lines that are drawn separating those who will and those who will not be prioritised for receipt of genetic medicine. Paradoxically their data also indicate that those deemed not to need clinical monitoring for genetic disorder (because they fall on the less risky side of the boundary) are those most likely to experience higher levels of anxiety. The classification and representation of genetic risk via genetic testing thus has both intended and unintended effects that will, writ large, have major implications for not only the structuring of health care but also how people 'make sense' of diagnoses/prognoses. Hundt et al. continue this theme in their analysis of prenatal screening for genetic disorders. They show how women bring to the clinical encounter diverse forms of knowledge and experience through which the clinical risk of carrying a child with a genetic disorder is framed. This framing of genetics according to a wide range of social criteria is a theme underpinning Kerr and Franklin's analysis of the ambivalence and ambiguities associated with genetic information and its meaning and value. Crucially, ambiguity need not be seen solely in negative terms, but rather provides a space through which lay and expert can open up a dialogue about the meaning of, say, an 'abnormal' pregnancy. Uncertainty is better embraced than, misleadingly or falsely, disposed of.

Part 2, *Information and Empowerment*, switches our attention to developments in medicine and health care associated with information and communication technologies (ICTs). Nettleton and Hanlon's chapter discusses the extent to which Internet and telephonic (NHSDirect) systems have changed the patient's 'pathway' to the doctor. While their data indicate that there have been some important shifts in information seeking and doctor–patient relations, they challenge the more radical claims made in recent medical sociology that these systems are corrosive of trust in medical expertise. Rather, they suggest that many of the arguments advanced 20–30 years ago remain valid today. This illustrates very clearly why we must avoid any simple technological determinism: the social relationships associated with trust-seeking are still of central importance in medicine. The ways in which information sources, knowledge, and trust relate to each other and are mediated by different technologies is explored further in the chapter by Green et al. which examines women's experience of and response to hormonal changes in midlife. Again, the findings show how, in clinical and non-clinical settings, women respond to such changes through reference to the wider context of their personal lives.

There have been many claims by health policymakers that telemedicine will radically transform the diagnosis, management and delivery of medicine. While there have been large investments in this technology throughout the world, the chapter by Finch et al. shows how such investments often neglect to consider the cultural, organisational and professional 'work' that needs to be done to 'normalise' and make telemedicine or telecare useful. Moreover, the presumption that patients will happily take up and be 'empowered' by what we might call 'distance medicine' as though this were thereby patient- rather than expert-centred is challenged by the authors. This again reiterates a theme in many chapters of the book that we need to be extremely careful about claims that new technologies are a) patient-centric and, thereby, b) empowering of patients. Ward et al. explore patient expertise as an expression of the embodied experience of illness and how, despite pressures from established medicine, this can be mobilised to make sense of and even *redefine* different disorders.

This question of empowerment is a key focus of the chapter by Cullen and Cohn in their studies of 'collaborative knowledge systems' and brain imaging in neuro-psychiatry. They argue that 'technical codes' – the dominant and stable design and function of systems – are always subject to appropriation by users in ways quite unexpected by those developing them. They argue that technologies need to be open to and shaped by the 'communicative practices' of their users, and that this begins from their very inception. Innovation is thereby simultaneously social and technical, and it is this point that marks out the chapters of Part 3.

Part 3, *Innovation: Context and Meaning*, brings together a number of chapters that examine the dynamics of innovation that shape health technologies. Many of these processes are found elsewhere, generic to all technological development. Metcalfe and Pickstone provide detailed accounts of the technical, professional, clinical, regulatory and commercial dynamics that shaped the development of intraocular lenses in the eye and artificial hip joints. The authors show how the 'design space' of each has been populated by diverse and competing interests, such that while there has been some closure around a popular design in each case, this is never final, with new possibilities and solutions to (both old and new) problems arising. Parr, Watson and Woods explore how innovation in design with respect to the Internet and the wheelchair – again strikingly distinct domains – betrays very similar processes, their focus here on the way in which both are 'sites of political conflict'. One of their key points is that users of the Web or the wheelchair are often 'invisible' to the design process: in their efforts to make themselves 'visible', users have to become more directly involved in design and thereby renegotiate the meaning of their disability in order to normalise it.

Seymour et al. provide us with an account of how innovative health technologies are deployed in new ways (and new places) at different times in the life course, from birth, through childhood to terminal care. As we saw earlier, the themes of ambivalence and risk and the tension between technologies enabling of greater control set against new forms of constraint are central to their argument. They show how these technologies can be 'out of time and place' if designers and facilitators fail to understand the wider context in which they will used. This is an issue that those who regulate and evaluate IHTs would surely understand, but how they do and how far they do so is the topic of the final section of the book.

Part 4, *Regulation and Evaluation of IHTs*, encompasses a range of very different substantive fields, from the use of the Internet, drug development, through tissue engineering and xenotransplantation, alternative medicine to embryo research; each chapter providing an opportunity to explore the relationship between the material/biological aspects of innovation and the governance and evaluative questions this throws up.

Abraham et al. track the relation between the material properties of drug compounds, innovation and their regulatory oversight. Both drug hype and drug licensing are open to challenge and critique. Brown et al. discuss how regulatory agencies have sought to handle new 'hybrid' technologies – here tissue engineering and xenotransplantation: these challenge the neat boundaries and categories of science and innovation that agencies conventionally use to regulate technology. Regulation is also anchored in and expresses a wider range of social and cultural values: standards of and restrictions on technoscience encode these values, more or less explicitly. Salter shows how diverse cultural values inform, structure and help explain different regulatory regimes in regard to embryo research and the 'continuing volatility' of European politics, while Chatwin and Tovey describe how alternative medicine is evaluated and indirectly 'regulated' through organisational and professional controls set by orthodox medicine.

Finally, continuing the theme that the evaluation of IHTs is socially embedded, Armstrong shows how the very metrics used to determine the quality of life (QoL) encode diverse and contestible social values. He shows too that these metrics play a central role in how all technologies – such as new interventions and devices – develop and are deemed to be successful. One of the principal points he makes therefore is that QoL measures are both a technology in themselves while acting as a medium through which other IHTs must pass before they are deemed to 'work'.

Conclusion

This brief introduction has only scratched the surface of what is a rich and critical analysis of innovative health technologies. Most of the chapters refer to additional papers that have been or will be published by the authors that provide detailed accounts of the range of IHTs covered in the book. What we have tried to do in this volume is to pull together more generic issues that cut across discrete fields of medicine, clinical practice and health precisely because at their core we can find common processes at work. I have suggested earlier in this chapter four characteristics that mark out IHTs as being embedded, projective, hybridised and representational. Patients, practitioners, innovators and those who regulate innovation have to cope with the new demands these create: it is the aim of the work reported here that social science can contribute towards what Collins and Evans (2003: 448) have called an 'interactional expertise' to help better understand and manage these demands.

Part 1
Genetic Risk, Reproduction and Identity

Part 1

Genetic Risk, Reproduction and Identity

1
The Genetic Iceberg: Risk and Uncertainty

Aditya Bharadwaj, Lindsay Prior, Paul Atkinson, Angus Clarke and Mark Worwood

Introduction

New genetic technologies make it possible to screen populations in order to identify individuals' risks or susceptibilities to specific medical conditions. On the basis of such population screening, otherwise healthy, symptomless people can be placed in the medically ambiguous position of being at risk. Equally, individuals may be tested on the basis of a family history that suggests a risk of an inherited condition. The identification of risk is probabilistic: it rests on the allocation of risk values. These are, in part, dependent on current understanding of the relationship between a given genetic make-up (the genotype) and its manifestation in physical characteristics or medical conditions (the phenotype). These are not always established with any degree of certainty and on the basis of genetic testing and family history, individuals may be identified as having high, moderate or low risks, and more specific values attached in percentage terms. Indeed, levels of risk assessment can appear to be arbitrary and moveable depending on how clinics choose to set the boundaries between, say, 'high' and 'moderate' risks. In some clinical contexts, such allocations to categories of risk are contingent upon organizational decisions and constraints concerning the allocation of resources.

Risk therefore is a complex category. In the context of population screening technologies, the identification of risk has potentially huge implications for the possible creation of risky populations. Following the notions of the 'clinical iceberg' (Last, 1963) and the 'symptom iceberg' (Hannay, 1979), we conceptualise the emergence of genetically risky subjects as the creation of a potential *genetic iceberg*. That is to say there is a pyramid of potential pathology (or potential future manifestation of disease) that lies beneath the relatively small numbers of people currently dealt with by clinical genetic services. This is not on account of an existing disease burden due to genetic causes, but rather because the criteria used to capture and define genotypic and phenotypic individuals at 'risk' is malleable – allowing for relocation of boundaries to include or exclude individuals or population groups from active screening. Such decisions we argue are centrally entrenched in the political calculus of health planning. They are informed by issues of resource availability and allocation, as well as by 'science', and the boundaries of ethical and practical decision making. The chapter explains how, in the context of genetic medicine, populations get categorised and how their screening, surveillance and therapeutic interventions get prioritised. There is, moreover, an iceberg of potential anxiety and uncertainty on the part of those identified as being at low to moderate risk, who are not allocated to active healthcare management (because their risks are comparatively low, in comparison with high-priority categories). The expansion of genetic testing may therefore create an unmet need for medical monitoring, counselling or – at least – reassurance.

Our discussion in this chapter draws on findings from two complementary research projects examining risk and uncertainty in the context of genetic haemochromatosis and cancer genetics. The two projects – and the medical conditions they examine – allow us to look beyond single types of genetic condition, in order to assess the possible range of consequences of genetic testing, while identifying some features that are common to them.

Haemochromatosis and cancer genetics: the case studies

Genetic haemochromatosis has acquired clinical and scientific significance since the discovery of the *HFE* gene in 1996 (Beutler et al., 2002). Haemochromatosis is a genetic disorder causing the body to absorb an excessive amount of iron from the diet (Bothwell and MacPhail, 1998). The excess iron is subsequently deposited in multiple organs, especially the liver, pancreas, heart, endocrine glands and joints and cause serious damage (Niederau et al., 1996; Bothwell and MacPhail, 1998). The susceptibility to absorb excessive amounts of iron is usually associated with homozygosity for a particular mutation of the *HFE* gene. Since its discovery in 1996, this mutation (the *C282Y* mutation of the *HFE* gene) has been identified as the underlying cause of haemochromatosis in over 80 per cent of the patients (Beutler et al., 2002). It is estimated that in Europe, Australia and the USA 60–100 per cent of patients with genetic haemochromatosis are homozygous for this mutation (Worwood, 1999).

As for cancer genetics it has been clear for some time that with some of the commoner cancers – breast, ovarian and colon – there is a Mendelian subset where genetic risk is estimated to be high, but which cannot be clearly distinguished on clinical or pathological grounds from other cancers. (However, gene expression methods are currently being developed to overcome this difficulty.) In the case of ovarian, breast and colorectal cancer the subsets can be identified by the use of molecular tests. For example the two genes associated with breast cancer (*BRCA1* and *2* – located on chromosomes 17q and 13q respectively) are believed to be carried by around 1.7 per thousand people – though they are probably responsible for only some 5 per cent of all breast cancers (Department of Health, 1998a). *BRCA1* is also associated with ovarian cancer. The incidence of cancer with a familial predisposition as a whole is probably in the region of 1 in 30 individuals – so about 10 per cent of all cancers have a familial subgroup (McPherson et al., 1994). The penetrance of some of these genes is estimated to be as high as 0.8 – implying that an individual possessing the mutation is at a very high mathematical risk of succumbing to cancer.

Our discussion of genetic haemochromatosis (GH) reports research on individuals identified as either affected by, or at risk of, iron overload from genetic haemochromatosis in three regional centres in England. The research project on haemochromatosis compared those individuals presenting symptoms of the disease as a result of the *HFE* mutation with healthy blood donors who through screening were shown to be potentially susceptible to the condition, carrying two copies of the *C282Y* mutation (homozygous), but as yet having no clinical manifestations of the disease. The broader aim of this research is to contribute to the ongoing debate on the desirability of population screening for genetic diseases (Clark, 1995; Burke et al., 1998; Allen and Williamson, 2000; Beutler et al., 2002).

This chapter draws only on the accounts of twenty people in the South Wales area who had tested positive for the mutation causing haemochromatosis but were otherwise healthy. Unlike those at risk of cancer, the haemochromatosis series were not allocated risk *values*. Because the relationship between inherited susceptibility and onset of clinical disease is uncertain, such individuals are identified as being susceptible to the condition, but their risks are not calculated as being 'high' or 'low'. The research focused on their accounts of the impact of the screening, and their personal constructions and understanding of risk/ susceptibility; their assessments of health care and support; and their personal experience of living with the risk of disease. We compared these accounts with those who had developed clinical haemochromatosis. The latter had presented to healthcare professionals with a variety of symptoms, and had been diagnosed not on the basis of genetic testing, but on their levels of iron. The Cancer Genetics project was concerned to discover how 'risk' estimates were assembled, discussed, made visible, applied and understood by various parties. They included medical geneticists, lab scientists, nurse-counsellors, and the people who came in contact with the genetics service. For the purposes of this chapter we draw only on interview data with what might be called the 'patients' of the service – though as we shall show, the use of that term is highly problematic. The sample of patients studied was stratified by level of risk and type of cancer and the sample composition is as indicated in Table 1.1. Before we look at these accounts, it is important to understand in broad terms how genetic risk is socially constructed.

Table 1.1: Sample of 'patients' interviewed in the Cancer Genetics Study

Cancer Site/by Risk	Lo risk	Mod risk	Hi risk	TOTAL
Breast	8	11	6	25
Ovarian	2	3	6	11
Colorectal	6	10	4	20
Other cancers	1	0	1	2
TOTAL	17	24	17	58

Screening, uncertainty and risk

Risk looms large in contemporary medical discourse (Gabe, 1995). Indeed, Skolbekken (1995) has spoken of a 'risk epidemic' and has documented the marked rise of risk-related publications in medical journals during the 1967–91 period. So powerful has been the impact of risk analysis that Gifford (1986) has argued that risk has now replaced 'cause' as a focal point for clinical research. More recently, Førde (1998) has viewed such developments as an expression of a specific risk culture – a culture in part bolstered and fostered by epidemiology and public health sciences, sciences that have been more concerned with aspects of probabilistic modelling than with microbiology per se. Those who adopt what is often referred to as the scientific position on risk claim that risk ought to be defined and assessed in terms of the concept of probability. Combined with this claim is another to the effect that scientific assessments of risk are both objective and factual (see, e.g. Royal Society, 1992: 4; 1997: 46) in a way in which lay assessments are not.

One problem with this line of argument is that 'probability' is a slippery concept. As the case of haemochromatosis demonstrates, risk assessments based on uncertain predictive technologies at best allow for the articulation of potentially risky outcomes in place of more conventional quantifiable, mathematical probabilistic risk projections. Tentative predictive technologies tend to verbalise risk as possibility rather than quantify 'risk as probability'. This displacement of 'numbers by words', while in no way peculiar to the case of haemochromatosis, mirrors a larger conceptual issue of what constitutes risk (Bharadwaj, 2002: 228). Indeed, within mathematics itself there are competing versions as to what 'probability' is (Hacking, 1990). It is not appropriate to outline alternatives here. However, it is appropriate to state that one of the most persuasive of mathematical approaches is that which equates a probability to the frequency with which an event occurs in a long series of events (Edwards et al., 2005). (For example, the frequency with which breast cancers are detected per year in a national population of females.) Clearly, it is such population-based estimates of probability that epidemiologists use in their analyses of the rates with which particular forms of disease event (influenza, meningitis, lung cancer) appear. Yet, whilst epidemiologists derive their risk estimates from population assessments, clinicians deal only with individual cases. Inevitably, in the interstices between populations and individuals various difficulties can arise.

For example, let us assume that a group of clinicians use the probability of 0.25 as a cut-off point for some cancer screening service or other. By implication, such a policy will be detrimental to the small proportion of people whose chances of contracting a cancer are 1, but who are 'buried' in a sub-group that has a very low (population) prevalence – say, middle-aged males with breast cancer. This would be so, of course, no matter what the cut-off point was to be. For what the probability assessment is based on is a group, or a collective, and not an individual. Naturally, having derived a probability assessment from a collective, practitioners have necessarily to apply the assessment to individuals. The application of population-based assessments to individuals involves what we might call a translation – though it is not the only one. For, as Gifford (1986) indicates, a second translation occurs at the point where the clinician treats 'risk' as a pathology. That is, at the point when he or she identifies 'being-at-risk' as a clinically manageable condition. Both translations are evident in the work of clinics dealing with adult onset genetic disorders.

The problem for such probability assessments is that it is not immediately clear who is most likely to develop a cancer that is related to an inherited deficiency in the absence of any signs or symptoms of disease. In order to determine members of such groups, initial selection criteria are used. For example, as far as breast cancer is concerned, initial inclusion in a 'higher than population risk' group requires at least one of the following criteria to be met. One first-degree female aged under 40 affected; 2 first-degree relatives (on the same side of the family) affected at 60 years of age or less, 3 first- or second-degree relatives of any age (on the same side of the family); 1 first-degree female with bilateral breast cancer; 1 first-degree male breast cancer. Different criteria (the Amsterdam criteria) would be used for colorectal cancers (see Vasen et al., 1991). For ovarian cancers the requirement would be for 2 or more such cancers within the family with at least one first-degree relative affected. (A first-degree relative would be a parent or a sibling.)

Similarly while in Britain 1 in 200 people are susceptible homozygotes for haemochromatosis (i.e. carry two copies of the mutation), the proportion who develop clinical haemochromatosis is small but not clearly determined. Unlike cancer genetics (as shown above) there is no clear-cut selection criterion present in the clinical domain. Therefore, while it is possible to screen individuals with the view to predicting their personal susceptibility to developing the condition, such risk prognostications are often tentative. A positive test result demonstrating two copies of the mutant gene identifies an otherwise healthy individual only as susceptible to the development of the disorder. A recent study in South Wales showed only 1 per cent of adult homozygotes had a clinical diagnosis of iron overload (McCune et al. 2002). There is as yet no firm clinical basis for the prediction of when and how a healthy susceptible individual develops 'frank disease' or overt iron overload, although it is becoming clear that variation at other genetic loci as well as environmental factors including diet are all relevant. Measurement of body iron stores serves as a proxy indicator – a very imperfect predictor – of future disease onset. This leaves clinicians and scientists still grappling with the multifactorial complexities underlying the condition and what counts as 'at-risk status'. If identified, the condition can normally be managed through regular bleeding, which depletes the body's excess iron, and the condition is eminently treatable (Niederau et al., 1996). If it remains undiagnosed, serious organ damage can result (see McDonnell et al. (1999) for a survey of 2,851 patients' experiences and symptoms). In the absence of any generic selection criterion that attaches probabilistic risk assessments to individuals a continual follow up on the individual / population is the only viable alternative.

However, the mere presence of selection criteria, as in the case of cancer genetics, is not sufficient. How such criteria are actually used by those who refer people to a clinic is not at all clear (Wood, Prior et al., 2002). What is important from our standpoint is that the boundaries between high and low risk implied by the criteria remain malleable. So it would be possible, for example, for the clinical service to alter the above filtering rules for cancer genetics so as to abolish, say, the age barrier of 40 in the first-degree relative, or to alter the requirement for three first-degree relatives at any age, to two such relatives. Shifting the boundaries in such a manner could thereby increase or decrease the population 'at risk' considerably.

Naturally, consequent to an initial identification of someone being at risk as described above, there would follow detailed investigation of the case so as to determine how probable it is that an individual might be a mutation carrier. Such investigations are based on detailed studies of family history and – where appropriate – of blood samples. In the case of cancer genetics such investigations have been described previously by Prior, Wood, Gray et al. (2002), and Wood, Prior, Gray et al. (2002). Of direct significance to this discussion is the fact that a person deemed to be at 'high' risk would not only gain from a detailed analysis of their chances of inheriting a mutation but would also be offered counselling and access to additional screening procedures and a system of health surveillance. A person deemed to be at moderate risk would also experience benefits of surveillance. However, people allocated to the most common category of risk – low or population risk – would not be encouraged to access any procedures additional to those available through primary care and therein lie some difficulties.

Embodying risk: the distressed and the indifferent

In the cancer genetic study of fifty-eight 'patients' who had been referred to a cancer genetics service it became clear that the ways in which individuals react to a risk assessment holds interesting lessons for the introduction of genetics and other new technologies into a health service. For example, when patient-clients reflected (via interview) on the experience of going through the genetics service, it became evident that many had struggled to comprehend their status in the health care system. Thus people often expressed uncertain feelings about the fact that they were not fully recognised as cancer patients whilst, on the other hand, were not perhaps entirely 'healthy'. Those who had received a genetic risk estimate as moderate or high often saw themselves as entering a 'liminal' state (Turner 1967), a state betwixt two worlds where they waited either to develop a disease and be treated accordingly or to be reassured from screening results that they were in good health. This uncertain status of being neither sick nor healthy introduced new concerns to the patient-clients as they attempted to integrate the risk estimates into their lifeworld. In particular they had to adjust to an increasing awareness that they might develop cancer in the coming years, but remained unsure as to when and indeed if this would happen. This period of anxiously waiting for test results, screening or firm diagnoses was identified by Crawford (1980) as the 'potential sick role' (c.f. Parsons, 1951). Crawford saw it as a position in which the integrity of the body is left hanging in question whilst the individual is denied access to the privileges of both the unwell and the healthy.

Interestingly, as far as screening is concerned, such outcomes had been predicted over two decades ago by commentators who noted how 'risk' rather than symptoms were beginning to form the focal point of clinical consultations (Brett, 1984; Gifford, 1986; Lefebvre, 1988). Indeed, in a somewhat percipient passage Davison et al. (1994: 355) argued that there was evidence from studies of hypertension that, 'asymptomatic risk identification can create a type of social identity (neither well nor ill but 'at-risk')' – and that this held enormous implications for predictive genetic screening where 'mass application could produce large numbers of well, but quasi-sick individuals and families'. Invoking the concept of medicalisation, Finkler (2001) similarly argues that the concept of inheritance of a predisposition can extend to convert most people into 'potential perpetual patients' causing people's realities to encompass a future 'incessantly punctuated by worry' (Finkler, 2001: 250). This state of being quasi-sick, liminal or potential perpetual patient to the health services is, in many ways, summed up in the following quotation.

P34: It's understandable that they need to see the higher risk people, rather than people like me ... But who am I?
FW: Well you are important.
P34: But I am not even a patient. That's another way of looking at it, I am not –
FW: Of course you are.
P34: I'm not, I'm not. I do not feel like a patient. Because to be a patient there has got to be something wrong with you. And there is nothing

wrong with me. I am just somebody who filled in a bit of paper. That is how I feel. And I would have felt like that even if I talked to somebody. I am still not a patient. (P34: Clinical assessment: moderate risk, colorectal cancer)

The challenges to self-identity and the lifeworld were perhaps most pronounced in those who were deemed low risk, for they were not usually invited to face-to-face consultations, but encountered the genetics service via telephone or written communications. For some, this created a sense of anti-climax. Patient-clients often felt that they had invested a great deal of emotional energy into completing a family history questionnaire and mentally preparing themselves for 'the worst' news, only to find that the outcome was deemed inconsequential, and they would not be given access to any additional screening or testing services. Thus some of the patient-clients who had been reassured that they did not need any further treatment interpreted this as evidence that they were not legitimate 'patients' and had been unfairly denied resources – especially resources relating to screening.

Often, people deemed to be at moderate or high risk of an inherited condition tended to reflect on that judgement with equanimity. Whilst, after being told that they were assessed as being at normal or population (low-risk) of inheriting a mutation, many patient-clients expressed a sense of dissatisfaction. (This was despite a concerted effort by the professionals in the clinic to explain the consequences of the 'good news').

In the haemochromatosis study most patients learnt of their asymptomatic mutation status through the post, followed it up and felt reasonably reassured once they got in touch with the clinic, as Jo, a 43 year old homemaker from South Wales reveals:

> ... I mean initially the letter was very self explanatory you know 'you needn't worry about this at the moment'. I think everybody is going to worry when they have a letter through the post saying well we have found this in you, you know. But when I went down to the hospital ... they allayed any fears that I might have had and tried their best to explain things to me. And even explained that I may never ever develop any symptoms from it or the need perhaps to be you know to go and donate blood more regularly or whatever, or to be bled. It might just be something I have to keep an eye on.

The information Jo receives in the clinic serves to allay her fears but equally shape her subsequent response. In not being able to predict the progression of the disease from asymptomatic carrier status to symptomatic case of iron overload the clinical realm can do little by way of contributing any definite information. The fact she may never develop any symptoms is but one possible scenario. For Jo to focus her thinking on the more positive outcome and not actively engage with other possible disease outcomes partly helps explain why the tentativeness serves to instil calm and on occasions even indifference amongst asymptomatic persons. Conversely, those who were categorised as 'high risk' within the realm of cancer genetics seemed to be quite content with the way in which the genetics service had configured their status. Indeed, the 'fortunate' outcome of receiving a high-risk estimate appeared to create feelings of safety, reassurance and trust in the power of medical knowledge. As patient-client 31 put it,

> P31: I'm glad the decision was taken out of my hands really. And I am pleased I am being monitored. I went for a mammogram today. (P31: Clinical assessment = high risk, ovarian/breast cancer)

Furthermore, in many cases, the understanding that other family members would also be eligible for screening and testing augmented these positive feelings.

The haemochromatosis study on the other hand found diametrically opposed assessment of low or no clear and present risk amongst healthy asymptomatic individuals testing positive for the *HFE* mutation. Jackie, a 42 year old college teacher based in South Wales had this to say:

> AB: Right, so are you glad you were screened for it?
> JS: I can't say I'm glad I was screened for it. It's blissful ignorance not to know. Because it's not affecting me in any way it's just another thing to have to think about, or at the time, I don't even think about it now. Er I can't say I'm glad, I suppose if it was affecting me and to have a name of a condition would be helpful. But given that I've got no symptoms I don't class it as helpful, but I don't class it as a hindrance either ... Well I don't think about it at all. The only time is when I get a letter from you lot to say A. can we come and screen you, B. can we screen your parents or your family or can you take part in a piece of research. It's the only time I think about it because I don't feel ill, I don't see anything. It's not like a cancer, if you've got cancer you know it's going to get worse. My understanding of haemochromatosis is as long as it's kept in check and I was told that checks would be made through my blood donor sessions.

In this case putting a name to the condition was enough for the respondent to carry on as 'normal'. The respondent above was not alone in expressing such sentiments since many felt that the low risk assessment made no difference to their everyday life. On the contrary they perceived no or little risk as 'being looked after'. In addition the comparison with cancer in the account above is particularly telling as it puts in perspective the sheer range of genetic based conditions and molecular genetic technologies of risk prognostications and the subsequent consequences for individual self assessments. Haemochromatosis is therefore not seen as life endangering by the asymptomatic compared to those people with familial genetic cancer. Thus the indifferent response to the condition is largely predicated on such a perception and, as we will see in the succeeding section, on the hope that someone is keeping an eye on them.

The desire to be recognised and taken seriously by the genetics service on the other hand is particularly evident in the case of cancer genetics patient-client 22, who was convinced that her mother's cancer signified a genetic predisposition in the family. Despite being told that she was at low risk of breast cancer, this patient-client remained determined to 'prove' her theory, and said that she was searching for more information about her family history and even considered obtaining a second opinion from a private practitioner.

This apparent paradox of treating 'good' news as 'bad' news can possibly be accounted for in a number of ways. For example, it could be that the receipt of a risk assessment that ran counter to a personal risk assessment was sufficient in itself to generate a sense of discordance. Indeed, Renner (2004), in relation to a study of feedback about cholesterol levels, suggested that information that is inconsistent with pre-existing expectancies is generally perceived as less trustworthy and accurate than is consistent information. It could also be that as 'genetic' causes of disease are commonly nested in a wider framework of aetiology, discounting the 'genetic' fails to fully discount the 'inheritance' – as is argued by Michie et al. (2003).

Both explanations find some support in the cancer genetics. However, we would argue that there is more than mere lay 'belief' involved in these paradoxical observations. Indeed, we suggest that what people are reflecting on in our interviews is not simply their health beliefs, but their position in the health care system. Consequently, to fully comprehend their reactions we need to consider the ways in which people are connected to the material structures in terms of which they organise their health care.

Explaining risk: 'keeping an eye on things'

As patient-client 41 explained, upon hearing the news that he had a significantly increased risk of colorectal cancer and was entitled to colonoscopies until the age of 60:

> P41: This [holds risk assessment letter from the clinic] is in the background. Am I going to get to 60? Let's worry about it when I get to it. You say what do I think about it? Yeah I am glad. I know that I can say 'Hey, I want some screening now.' And they have to do it too. Whereas if I go along and say 'my dad and grand-dad died of cancer can I have some screening?' Unless I had a genetic risk they would have said 'on your bike, back of the queue'. And at least now this is in my notes and when my time comes at least I know. (P41: Clinical assessment = moderate risk, colorectal cancer)

Patient-client 06 illustrated a similar outlook by proclaiming that while she relied more upon subjective perceptions and her lay 'expertise' than genetic knowledge to identify changes in her body, she would welcome the additional support of screening and monitoring to complement her own routines and practices.

> P06: I wasn't particularly worried about the genetic aspect of it, because as far as I am concerned I am more at risk anyway. I don't need a gene to tell me that. You know the fact that two sisters have had it, even if statistically I am not, in my mind I am more at risk. But I wasn't looking for a gene test to say I had the gene or anything. It was just really I wanted increased supervision ... I think quite honestly I am more at risk so I will be more careful. Whatever the extra risk is, I will just be keeping my eye open a bit more, and examining a bit more. (P06: Clinical assessment = moderate risk, breast cancer)

These references to being monitored, cared for, looked out for, and being supervised, are liberally scattered throughout our interviews. Similarly the haemochromatosis respondents expressed a particular 'faith' in being coded, monitored and looked after. To cite Jackie again:

AB: So you are being monitored from time to time?
JW: Whether I am or not I don't know but in my little mind I'm happy knowing that Big Brother is out there checking my blood for me when I'm not looking ...

Jo on the other hand was in favour of proactive screening to know what the future might hold. For her a category, label and a proactive stance was a responsible and sensible way forward:

> ... do as much screening as you can and just be aware of it. If you're aware of it and don't bury your head in the sand then the likelihood is that the family risk could even save your life rather than endanger it if you're aware of it isn't it really. Because if you're aware of it you'll come under a category, if it's a high risk one then you're far more likely to do your own screening and test and make sure that if there's something you're unsure of you get it checked out. Whereas I think if you don't come under that you just go through life thinking that won't happen to me and then by the time you do something about it, it could be too late.

A similar desire for surveillance is expressed by the cancer genetics individuals who hope for an engagement with a health care system that will constantly check and 'keep an eye on' the patient-client. In that sense patient-clients are not simply giving voice to 'representations' or health beliefs, but seeking ways of organising health care resources around their specific and particular needs and lifeworlds.

More importantly, perhaps, in trying to position themselves within a system of health care resources it could be argued that patient-clients are showing an awareness of the multifaceted nature of risk assessments. For – as we have pointed out – the latter are not simply objective scientific assessments of the probability of coming to harm, but are complex vehicles that determine the distribution of health care resources. In a sense they serve as rationing or distributive mechanisms in a health service system. Thus, users of the service are well aware, for example, that regular screening, counselling and other features of health care management are only available to those assessed as high and moderate risk. And as our extracts from the interviews (above) indicate, patient-clients are 'not worried' about risk in the sense of carrying a mutation. Their eyes are on the resources that follow the risk, and they see their task as positioning themselves so as to gain access to those resources. It is a stance that is clearly evident from the following extract – drawn from a consultation between a genetic nurse-counsellor and a service user.

NC2: OK. And do you want me to give you ... you know we are talking about high risk. Do you want me to give you any more details than that?
P31: No, 'high risk' is enough. That is the reason I am here.
NC2: OK.
P31: I just want there to be somebody looking after me. That is how I felt when [the clinic] finished. I thought 'I have been abandoned.'

Similarly, the predominant assumption in accounts of our haemochromatosis cases rested on the notion of being tracked and followed by 'Big Brother' or the health service 'keeping an eye' on them.

Conclusion

The question of risk in relation to health is one that has become increasingly important for both health professionals and medical sociologists. This is, perhaps, not so surprising in what has been termed the risk society (Beck, 1992), where – through our persistent attempts to colonise the future – 'the influence of distant happenings on proximate events and on intimacies of the self becomes more and more commonplace' (Giddens, 1991: 4). It is certainly the case that in modern medicine the focus on (and management of) risk status rather than signs and symptoms of actual bodily pathology has had considerable implications for patients, health professionals and for health provision in general. Above all, we have been concerned to demonstrate how the status of 'being-at-risk' can on the one hand position the patient or client of a clinical service in a liminal world betwixt health and illness and on the other hand generate insouciant, indifferent responses. This shows the wide range of conditions and clinical encounters that make up the experience of at-risk status and it has implications for the conceptual architecture of medical sociology and anthropology – most notably in so far as it reflects on the notions of 'health', 'illness' and the sick role. For instance we have suggested that within the ambit of cancer genetics the recognition of liminal status seems to be sharpest among those deemed to be 'low-risk' (in that sense, gradations of risk status are significant). And we have noted how such low-risk patients or clients express a sense of dissatisfaction with their risk status and sometimes strive to reposition themselves into higher risk categories. On the other hand we have demonstrated how low risk is almost always interpreted as no risk amongst asymptomatic haemochromatosis homozygotes.

There are various attempts to conceptualise the shift from 'healthy' to 'as yet not ill'. They range from the notion of 'predicting diseases before symptoms appear' (Nelkin and Tancredi, 1994) and 'people as patients before their time' (Jacob, 1998) to 'the asymptomatically ill' (cf. Novas and Rose, 2000) and 'potential, perpetual patients' (Finkler, 2001). There is, however, little critical focus on those who do not experience being at risk in those terms. These conceptualisations, and the findings from the cancer genetics research reported here amply demonstrate that molecular genetics can produce a heightened sense of anxiety and uncertainty in individuals identified as 'at risk'. Nevertheless, we must be careful not to over-generalise or over-emphasise the possible creation of 'risky populations'. As the data from the haemochromatosis research has shown, there is variability in people's risk perceptions. It is not the case that the informants who had been identified as being susceptible to haemochromatosis necessarily experienced heightened anxiety. There is an important difference between those at risk of cancer and those susceptible to haemochromatosis. The latter does not carry the same personal threat or the same cultural significance as cancer, serious though it can be. Moreover, in the absence of prior family history, population screening for a condition like haemochromatosis may not have the same experiential value for those potentially affected. We do not assume, therefore, that any or all modes of genetic testing will necessarily lead to the creation of a uniformly 'risky' population, at least in terms of lay people's own phenomenology of health and illness. These are empirical questions that need to be explored systematically through further investigations of a range of testing procedures and a variety of medical conditions. Overall, however, it is clear that the deployment of new medical technologies – such as those of modern clinical genetics – serve to redefine the status of health and of illness. In so doing innovative health technologies draw into the orbit of medical care a new population of people who are not yet ill, but who are seen as having the potential of succumbing to some identifiable disorder.

That process in turn leaves at the margins an even larger population of healthy individuals – either anxious about their state of health, feeling puzzled or living with a (possibly false) sense of security that either someone is or is not 'keeping an eye on them'.

Acknowledgements

We acknowledge the financial support of the Economic and Social Research Council (grant nos. L218252009 and L218252046). These projects were part of the ESRC's Innovative Health Technologies Research Programme. We are grateful to Andrew Webster, the Programme Director, for his support and encouragement. The work was conducted under the aegis of the ESRC Centre for Economic and Social Aspects of Genomics (CESAGen), a collaboration between Lancaster and Cardiff Universities.

2
Navigating the Troubled Waters of Prenatal Testing Decisions

Gillian Lewando Hundt, Josephine Green, Jane Sandall, Janet Hirst, Shenaz Ahmed and Jenny Hewison

Introduction

'Informed decision making' is considered to be a basic prerequisite in prenatal testing, but there are many difficulties with this concept, not least in defining what information people need (Green et al., 2004). Parents need information about the pros and cons of accepting prenatal screening, as well as the specific processes involved. Additionally, if parents-to-be are to make decisions that are consistent with their own value systems, it is likely that they need some information about the target condition so that they can form a judgement about what it would be like to live with an affected child. This chapter brings together two studies within the Innovative Health Technologies Programme that focused on the social context and implications of prenatal genetic screening and testing technologies. One examined the social and organisational implications of innovative and established modes of prenatal screening as defined, perceived and communicated by health professionals and pregnant women. The other explored women's views about a range of conditions for which prenatal genetic screening is likely to be possible in the future. This chapter draws on the thoughts and feelings of women from both studies concerning the possibility of having a baby with one of the conditions for which there are or will be screening tests.

Previous studies show that although almost all women have heard of Down's syndrome, they are less likely to know medical details or detail about levels of functioning, which is indeed so variable (Mulvey and Wallace, 2000). Ethnicity, parity, religion, and knowing a child with Down's syndrome also affects women's knowledge (Chilaka et al., 2001), as does age and educational level (Green et al., 2004). As Press and Browner et al. (1997) indicate, the issue of disability becomes foregrounded when women are pregnant and being offered prenatal testing.

There are also concerns about the balance of the information and informed choice. There is an existing debate around the nature of the information given by health professionals to pregnant women about Down's syndrome. For example research suggests that the professionals giving this information often have no first hand knowledge of people with Down's Syndrome, and that they therefore tend to rely on medical textbook-type descriptions which focus on the potential problems of the condition. Also these health professionals lack time to talk with women and so tend to rely on information leaflets, which focus more on the screening process than on the condition, and which tend to lack positive statements in their brief descriptions of the condition (Williams et al., 2002a, 2000b).

There are concerns about whether pregnant women are being supported to make informed choices about screening for anomalies located within broader ethical and public policy debates concerning the rights of disabled people (Kerr and Shakespeare, 2002) and the need to be aware that these innovative technologies in the clinical setting have a technological imperative that, once routinised within the clinic, can limit choice through a risk escalator that can become difficult to get off (Kerr, 2004).

Rather than examine the quality of available information or the views of health professionals, this chapter explores women's knowledge and views concerning Down's syndrome and other conditions. The data from these two studies show us that women of all ages and background are aware of the complexities concerning prenatal screening and testing, give these issues considerable thought and have acquired knowledge from meetings with health professionals but also more significantly perhaps extensive experiential knowledge through their social networks of family and friends, their religious and spiritual beliefs, their previous pregnancies and their own reading and interaction with a range of media such as television and the Internet. The influence of the encounters with health professionals during pregnancy is not the only or indeed the most significant factor in shaping these views or decisions where experiential knowledge is combined with authoritative knowledge.

Study 1
The views of pregnant women concerning Down's syndrome screening who have experienced innovative and/or standard antenatal screening provision

Current policy is to offer screening for Down's syndrome routinely to all pregnant women in England and Wales (NICE 2003). The introduction for the first time of a *routine* offer of screening for Down's syndrome, and the move to first trimester combined screening technologies which achieve the greater level of accuracy required, have the potential to generate new problems for staff and patients in the delivery and provision of health care (Brown and Webster, 2004). Prenatal screening for Down's syndrome has largely developed as a consequence of advances in technology and the presumption that the NHS would provide it as a health related service. However, the development of prenatal screening technologies is a contested and politically charged arena with ethical and public policy considerations.

Prenatal screening for Down's syndrome is unusual as a medical service in that its value does not lie in managing or curing illness but instead in simply producing information that generates choices for women. As Press and Browner (1993) have shown, the message that 'the State says that you must be offered this test' can all too easily be translated into 'the State says that you must have this test' with a reduction, rather than an increase, in the amount of informed decision making that occurs.

There is considerable debate about the role of reproductive technologies and women's choice. It has been argued that women have found reproductive technology empowering (Davis-Floyd, 1994) and been able to make selective use of technology in very sophisticated ways (Lewando Hundt et al., 2001). However, increased choice has also been seen as a risk to women's health (Lippman, 1999) or as a vehicle for the new eugenics (Rapp, 1999). Allied to this are specific concerns about the provision of informed choice within the current UK maternity care context, which may be more a matter of (partially) informed compliance (Stapleton et al., 2002).

The study reported here took place in two district hospitals, which served geographically and demographically similar populations. One offering innovative first, and the other second trimester screening. Nuchal Translucency screening in combination with maternal serum screening was offered with results provided within one hour. The second site provided standard second trimester biochemical serum screening where results were returned within one week, and those at higher risk would be offered amniocentesis as a diagnostic test. Uptake at the innovative site was very high, with 97.5 per cent of women accepting screening (Spencer et al., 2003), compared to around 64 per cent of women in London accepting second trimester screening (Kennedy and Saunders, 2002).

Data in this study are drawn from: a postal questionnaire sent to 1,649 women who were between 23 and 30 weeks pregnant (response rate of 64 per cent); 45 observed sessions in the hospitals and community; semi-structured interviews with 45 health professionals and a cohort of 27 women and some partners on a range of screening pathways. This rich quantitative and qualitative data set yields some insights into how women are informed about Down's syndrome and how they think about giving birth or not giving birth to a child with Down's syndrome (Sandall et al., submitted, Lewando Hundt et al., 2001).

The main study findings (Table 2.1) were that:

- Despite being offered screening for fetal anomalies, 80 per cent of all women received no information about Down's syndrome or the implications of having a child with Down's syndrome. Young women and those with lower levels of education in particular said they wanted more information and women in these groups were also more anxious overall. More work needs to be done on meeting the information needs of these groups of women.

Slightly over half of the women in the survey (55 per cent) said that they made up their minds about whether to have screening before they were pregnant or prior to their booking appointment when screening was first offered. This underlines how this decision for many women and their partners is taken prior to being pregnant or attending the clinic and there is no significant difference in this between women who were pregnant for the first time or those who already had children.

Interestingly, there was a range of views concerning whether women felt they had received enough information about screening for problems with the baby from health professionals. Twenty-five per cent of those with low education and 25 per cent of those women with higher education felt that problems with the baby had not been fully discussed whilst this percentage was smaller for those with GCSEs or A Levels (17 per cent and 15 per cent respectively). This indicates that women have different information needs that cannot be met by a standard 'one size fits all' approach.

In general, women reported that more information was provided about the screening tests than was provided about Down's syndrome itself or what parenting a child with Down's syndrome would be like. In spite of this, over three-quarters of women felt they had received enough information in general. The need to focus on informing women about the screening tests within a short amount of time, seems to have meant that relatively little

Table 2.1: Genetic screening: Information provided to women

What screening tests are available	64 per cent
What abnormalities are screened for	51 per cent
Detection rate of screening	62 per cent
How the test is done	70 per cent
What the results mean	55 per cent
What the options after the test are	34 per cent
What is it like having a child with Down's syndrome	8 per cent

Table 2.2: Reported sources of information about Down's syndrome

Health professionals		Community resources	
Midwife	31 per cent	Magazines	21 per cent
Hospital doctor	6 per cent	Knowing someone with DS	20 per cent
GP	4 per cent	Friends/family	19 per cent
Health care asst	2 per cent	TV/radio/Internet	17 per cent
No sources of information from any source	28 per cent	Support/campaign group	3 per cent

information was given concerning what it is like having a child with Down's syndrome. However, it would be misleading to think that all women have no knowledge about Down's syndrome.

Although 28 per cent of the women reported having no sources of information about Down's syndrome, the remaining 72 per cent reported that they had a range of sources, often more than one (Table 2.2). For those who reported they had information from health professionals, midwives were clearly the most important source compared to other health professionals (31 per cent). Non-professional sources of information were reported most commonly as being magazines, knowing someone with Down's syndrome, through friends and family or the media.

Eleven statements were made about the possible implications of living with Down's syndrome, covering medical, social and educational factors. These were based on information from previous research in the area. For each statement, women were asked to say if it was true or false. The figures in bold (in Table 2.3 below) indicate the right answers. The results indicated that despite having received little information from health professionals on the topic, women had fairly good knowledge of the implications of Down's

Table 2.3: Perceptions of Down's syndrome

For each of the following statements, please say whether you think it is true or false	True	False	Don't know
People with Down's syndrome now live into their middle age	**78**	2	20
Some people with Down's syndrome can attend mainstream schools	**75**	4	22
Children with Down's syndrome will be incontinent until adulthood	2	**58**	40
Some people with Down's syndrome lead independent lives	**78**	5	17
All babies with Down's syndrome have heart problems at birth	4	**36**	59
People with Down's syndrome are more likely to develop cancer	1	**47**	52
With Down's syndrome, learning difficulties are *always* severe	6	**67**	27
People with Down's syndrome sometimes pass GCSEs	**62**	3	35
All people with Down's syndrome have some learning difficulties	**56**	12	32
With Down's syndrome, it's not really possible to hold down a paid job	5	**73**	21
People with Down's syndrome can learn to read and write	**90**	0	10

syndrome – or at least were able to make a good guess. For each statement, only a small proportion of women picked the wrong answer – although on the medical questions (for example the likelihood of heart problems, cancer or incontinence among people with Down's syndrome) they were more likely than with the other questions to say that they did not know.

About three-quarters of the sample knew that people with Down's syndrome live into middle age, that they can sometimes attend mainstream schools and hold down paid jobs. Nine-tenths knew that people with Down's syndrome can read and write. If you score women against the information provided by the Down's Syndrome Association, 12 per cent scored 0–3 out of 11, 34 per cent scored 4–7, and 54 per cent – over half the sample – scored 8–11 out of 11. The figures were very similar for women who had screening and those who had no screening – so there was no relationship between how well-informed women were and the decision they made about screening.

So in summary, although women receive little information from health professionals about Down's syndrome, many of them have fairly accurate knowledge. This seems to indicate that many women have both authoritative and experiential knowledge about Down's syndrome. However, there are still some women, who lack information and/or would like more, particularly those with very low levels of education and those with very high levels of education, although their needs are very different.

Women repeatedly talked about children or adults they knew with Down's syndrome and also their observation of the parents of these children. They knew them through their wider family and social networks, through meeting at schools or in the course of their working lives, and they build up experiential knowledge from these encounters.

Some were very aware that Trisomy 21 (Down's) was an anomaly that is less serious than Trisomy 13 or 18 where the child will not survive long and therefore the possible presence of Down's syndrome poses the parents with more of a dilemma.

> 'Well we went shopping to ... and we saw a little Down's baby and also one of the workers at ... has Downs – a lad of about 25. And you just think "well there's still Down's babies being born" and did that mother then have a choice or did she not know because there wasn't screening available? ... It made me feel glad that at least I didn't have that choice. At least it was going to be a fatal abnormality rather than a Down's.'
> (1.1.ii 2nd interview with a woman at 14 weeks gestation, who screened as higher risk T13/18, referred for further diagnostic testing CVS, outcome no anomaly)

Women often drew on their personal experiences with Down's children when making their decisions. Some, on the basis of personal experience of knowing others who were parenting a Down's child, had developed their own views about screening and testing and termination before they were pregnant and these varied.

> 'We'd already decided that if the scan did come back that it was a high risk, we would go for the tests and then go from there ... It was decided before I was pregnant ... We always used to discuss it because in my extended family ... he's Down's and I've watched, it's my cousin's child, and I've watched her really struggle.'
> *(1.15.i first pregnancy, first interview at 12.5 weeks, lower risk, degree)*

The same woman at a third interview after giving birth said,

> 'We had decided that if we had a very high risk of Down's syndrome we would have an abortion.'

A different view was expressed by another interviewee:

> 'Well I would have said that it's a child who has a slower learning capability than a child that is born without Down's syndrome. I've never sort of seen Down's syndrome as a problem. When I used to go horse riding, there was a girl there who was the lady's daughter. She had Down's syndrome and she could ride a horse better than anyone and sometimes she went into a tantrum, but when she wanted a cuddle, you know, would cuddle everyone, she was so lovely. And I think that was my sort of overall view of Down's syndrome children. It depends on what level they are. You know, they just need a lot more attention. They've got a slower learning capability than anyone else ... but I didn't realize that it was the chromosomes that made Down's syndrome. I thought it had to be in the family or something ... I think I was more worried and it sounds awful about the social outlook of everyone else ... It wasn't until I actually sat down and thought well would these things bother me? ... Then I thought "Well no they wouldn't" I don't see it as a problem. I definitely don't think it's one of the biggest things that could happen.'
> *(1.2.ii 2nd interview with a woman at 17 weeks gestation who was screened as higher risk of Trisomy 21 (Down's syndrome) and who refused the diagnostic test of an amniocentesis)*

Women often reflected in interviews on the issue of being 'selfish' or 'fair'. This was used in contrasting ways as either being selfish to terminate a pregnancy or to have a Down's syndrome child in terms of this not being fair on other children in the family.

> 'I'd say if I knew that the baby was gonna come out with Downs or spina bifida then I wouldn't have the baby cos for me personally I don't think it's fair on the baby nor on us either.'
> *(1.16.ii 2nd interview with a woman at 16 weeks gestation at time of interview)*

Partners sometimes seem to have had different views but women who were interviewed talked about discussing issues with their partners and 69 per cent reported that their partners encouraged them to have screening for Down's syndrome.

> 'I would have gone obviously ahead with having further tests. My husband and I did talk about it briefly at the beginning and I remember sort of saying – "oh well you know we'd never get rid of my baby." And then ... He's sometimes sort of a bit more practical and rational and he sort of said ..."Do you think that you'd be able, or do you think we'd be able to, you know, look after a Down's syndrome baby sufficiently and do you think it's fair?" '
> (2.4.iii 3rd interview with a woman after she had given birth, who had had private first trimester screening with a lower risk result)

Sometimes the idea of being fair was linked to having other children already.

> 'I personally just don't think I could cope, knowing what I'm like with our child and not being able to cope with working and our child now ... I mean it's probably a very selfish attitude to have.'
> (2.11 First interview with a woman at 9 weeks gestation who was screened as lower risk)

Religion was often referred to as an influence on decision making and views about possible termination. In the interviews, the effect of personal beliefs about religion or fate were expressed.

> 'I'm Catholic so I believe in a God and to be honest with you, I think that if God gives me that I going to have a Down's syndrome baby, I going to have to keep up with that. And I wouldn't destroy that life because of the baby's Down's syndrome I don't know about my husband you see. He has a different opinion.'
> (2.10.i First Interview with a woman at 7 weeks gestation, offered triple test, only had the scan)

Others were simply fatalistic without referring to God.

> 'I'd rather just let it be. Whatever's going to happen is going to happen and we'll deal with it. That's the way we're looking at it.'
> (2.2.i first interview with a woman who was 14 weeks pregnant in her first pregnancy and was having the triple test)

Whereas this first study focuses on the experiences of women and health professionals experiencing current innovative and standard technologies for prenatal screening for Down's syndrome, the second study examines women's views concerning a hypothetical future scenario of accurate prenatal screening tests for many conditions. It also provides information concerning the way in which women think hypothetically about screening and possible termination for different conditions.

Study 2
Similarities and differences in women's attitudes towards testing and termination for different conditions

Advances in genetics and related technologies mean that prenatal tests are becoming technically possible for a very wide range of conditions. Many of these conditions will be unfamiliar, and not all will be 'disorders' in everybody's eyes. We are not justified in assuming that people who want prenatal testing for say, Down's syndrome, will also want to be tested for cystic fibrosis or deafness. This creates a dilemma, since the constraints of a busy booking clinic will not allow time for individual discussion of all possible conditions that might be tested for.

If the option to be tested for some conditions, but not others, is desirable how is the consent procedure to be managed? One approach might be to seek consent to testing for a limited number of *categories* of problem, but this of course begs the question of how the categories should be formed. This was the starting point for this second project: to investigate the feasibility of devising just such a classification system. We did this by asking over 400 women who had recently had a baby about their attitudes towards testing and termination for a list of 30 conditions. Half the women were Pakistani and half were white British. Half of each group had a basic level of education and the other half had a higher level of education. All lived in the North of England.

We wanted to know, firstly, whether there were consistent clusterings of disorders which could form a basis for a grouped consent process, and secondly, if so, whether these were consistent across women with different levels of education and from different ethnic groups. We also wanted to understand why women felt the way that they did about different conditions, so 60 of the women were also interviewed, as were a further 19 women recruited through Genetics clinics who had first hand experience of disability.

In contrast to the study of current screening, ours deliberately asked women to make hypothetical decisions. Answers to hypothetical questions do not always correlate well with actual behaviour because so many other factors come into play in the 'real world'. For example, in current scenarios women have to weigh up the pros and cons of screening tests with limited sensitivity and specificity and of diagnostic tests which involve a risk of miscarrying a healthy baby. Decisions can also be influenced by the way that tests are presented. Hypothetical scenarios may not be the best way of determining actual test uptake, but they have the advantage of removing all these factors and allowing respondents to focus on their underlying values. This was the preferred methodology for this study because we wanted to understand people's feelings about the characteristics of different disorders. We chose to ask these questions of women who had recently had a baby, rather than those who were currently pregnant because we did not feel that it was ethical to confront pregnant women with 'choices' about tests that were not actually on offer. Women were asked to assume that the hypothetical test would be carried out early in pregnancy, using routinely collected blood, and that it would tell them definitely whether or not the baby had the condition under discussion. We also felt that it was important to give people a brief description of the characteristics of each condition, rather than its name, since many would be unfamiliar.

What we found was that, with regard to attitudes towards testing:

- Except for the minority of women who said that they would want testing for everything (25 per cent), or for nothing (3 per cent), most women did not have an attitude to prenatal testing that was independent of the condition being referred to. Most women said that they would have testing for some conditions but not others.
- Pakistani women wanted testing for more conditions than did white women.
- Independent of ethnicity, less educated women wanted testing for more conditions than their more educated counterparts.
- The less educated Pakistani sample differentiated least between conditions, and the more educated white sample differentiated the most.
- The four subgroups converged at the 'more serious' end of the spectrum.
- The great majority of women in all four subgroups wanted prenatal testing for four conditions: absence of a brain (anencephaly); severe learning difficulties with early death (trisomies 13 and 18); being unable to move from the neck down (quadriplegia), and progressive muscle-wasting with death probably before age 20 (Duchenne muscular dystrophy).

Results regarding attitudes towards termination were:

- Most women did not have an attitude to termination of pregnancy that was independent of the condition being referred to. Most women said that they would have a termination for at least one of the conditions on the list.
- Pakistani women would have a termination for fewer conditions than would white women.
- Education was not related to attitudes to termination of pregnancy in either white or Pakistani women.
- For most of the conditions, fewer than a quarter of participants would want a termination of pregnancy.
- The four subgroups were very similar in their attitudes for most of the conditions studied. For some conditions, attitudes to termination were more favourable in Pakistanis, for other conditions, they were more

favourable in white women. For the majority of conditions, attitudes did not differ between groups.
- For a few conditions, there were differences in attitudes between education subgroups within the white sample; on each occasion, less educated women had the more favourable attitudes to termination of pregnancy.
- Attitudes between subgroups diverged at the 'more severe' end of the spectrum.
- Despite the divergence, women from all four subgroups were most in favour of termination for the same four conditions noted above, i.e. anencephaly, etc.
- The groups diverged in respect of the condition which ranked fifth overall, severe learning difficulties with a normal lifespan: white women were much more likely to favour termination for this condition than were Pakistani women.

In addition, it was clear that many women did not see prenatal testing as necessarily leading to termination in the circumstances of an affected pregnancy. The difference between the two attitudes was particularly marked in Pakistani women.

In terms of our aim of finding a basis for grouping conditions for consent purposes, the only consensus was with regard to the four conditions judged most severe: anencephaly, trisomy 13 or 18, quadriplegia, and Duchenne muscular dystrophy. Attitudes to other conditions were too individual to give a basis for categorisation.

What reasons did women give for their attitudes?

Whatever a woman's feelings about particular conditions, the same categories of reasons tended to be invoked to explain her choices. Women spoke of 'serious' and 'less serious' conditions. The four conditions mentioned above were generally agreed to be 'serious', but there was much less agreement about what was 'not serious'.

Women generally referred to the child's likely quality of life:

> '… cannot do anything, can't eat or ask their parents for anything'
> *(Pakistani, lower education, 4 children, recruited via Genetics clinic)*
> '… are severely brain damaged … totally paralysed'
> *(White, lower education, 2 children, recruited via Genetics clinic)*
> '… would die soon after birth or during childhood'
> *(Pakistani, lower education, 2 children)*
> '… are going to be suffering, in pain …'
> *(White, higher education, 2 children)*
> '… couldn't enjoy their life … understand what they're living or why …'
> *(Pakistani, higher education, 1 child)*

They also spoke about the quality of life for themselves and their family:

> 'I know I couldn't cope, I'd go mad myself'
> *(Pakistani, higher education, 2 children)*
> 'It's purely selfish but I don't know whether I could ... put 24 hours care into a child for the rest of my life'
> *(White, higher education, 1 child)*
> '... not fair on the first child ... would put a strain on absolutely everybody'
> *(White, higher education, 1 child)*
> '... My other children would feel left out'
> *(Pakistani, lower education, 2 children)*

The two conditions in which the baby would have a very short life illuminated some differences between the white and Pakistani women. These were the only two conditions in which Pakistani women were less interested in prenatal testing than white women. When talking about reasons *not* to have a termination of pregnancy, both Pakistani and white women invoked the concept of 'normality':

> *Blind or deaf* – 'they're still normal ... know what's going on'
> *(Pakistani)*
> *Duchenne Muscular Dystrophy* – 'Muscle wasting doesn't affect the brain ... so they are a normal person'
> *(Pakistani, higher education, 1 child)*
> 'End up in a wheelchair, but can laugh, go out with friends, and do things for themselves'
> *(White, higher education, 1 child)*

Pakistani women were more likely to mention religion as a reason for not having a termination, but most thought that serious conditions justified making an exception to this rule. There are varied interpretations of the Shariah (Islamic Law). It is considered that there is nothing wrong with a pregnant woman undergoing medical tests, as these can contribute to the health of the child as well as the mother although there are different rulings (fatwas) on the grounds and timing of the termination of a pregnancy.

> 'If you go by Islam, Islam says don't have a termination of pregnancy ... Allah has given it to you and you should accept it.'
> *(Pakistani, lower education, 5 children)*
> 'I'd be scared to have it terminated ... it's a big sin.'
> *(Pakistani, higher education, 2 children)*

But both of these women said 'yes' to termination for the more severe conditions.

> 'I'd rather go to hell than put my child through all that.'
> *(Pakistani, higher education, 4 children)*

The possibility of family and community disapproval was apparently not an important factor for women in either group. Pakistani women were more likely to refer to gender issues, specifically the marriage prospects of affected girls and the impropriety of a boy past puberty needing his mother's assistance with intimate functions. Women did not always specify where their appraisals of a given condition came from, but inevitably there were references to friends, neighbours and the media as sources of information. These single case examples seemed to be quite influential to women's thinking; they were rarely aware of whether their example was typical. Examples from the media generally seemed to be in the context of cures and treatments now available, which was an important consideration in many women's reasoning. Those women who had a child (or had lost a child) with a genetic disorder advanced very similar reasons for their views to other women. Overall, they tended to hold more favourable attitudes to termination.

In summary, this study indicated that there was widespread agreement concerning conditions that were considered to be 'serious' but much less agreement concerning screening for, or termination of, 'less serious' conditions. It was only possible to group four conditions and attitudes to the others were more individual. We concluded that a consent process based on groupings of disorders was unlikely to work because individuals have such different appraisals of what characteristics of different conditions seriously impair the quality of life of the child and the family. These differing appraisals arose partly from different value systems but also from different exposure to the conditions under discussion. Where does this leave health professionals in dealing with informed decision making as tests for more conditions become available?

Discussion

Study 1 shows us that women are not informed about the implications of having a child with Down's syndrome by health professionals. However their knowledge gained from other sources is high and their personal experience of knowing children with Down's syndrome seems to inform their views or the way they explain their views. It also shows that in the current system of prenatal screening health professionals struggle with covering what needs to be discussed. The second study looked for patterns in different women's attitudes towards a wide range of conditions. The only consensus found was at the severe end of the spectrum; apart from this it was not possible to group conditions together as a basis for a 'grouped consent' procedure.

The starting point for Study 2 was that, in the future, women will be placed in a situation where they could be tested for a very wide range of conditions. We tend to overlook the fact that this is not just a scenario of the future: it is already happening. In Study 1 women accepting a test for Down's syndrome may, especially through ultrasound screening be given a diagnosis of a vast range of other anomalies – chromosomal, cardiac and neurological problems. In regards to the first of these, Down's syndrome accounts for only about half the chromosomal anomalies detected at amniocentesis. Many of these other chromosomal abnormalities, such as trisomy 13 and 18, fall into the category that most women agree is 'severe' and that most women therefore would want to know about. However, not all chromosomal abnormalities fall into this category. Conditions like Kleinfelter's syndrome and other sex chromosomal anomalies confront parents with information that is much more difficult to deal with and which many would probably rather never have known about; moreover, professionals have been shown to provide confused and misleading information about the results of such scans (Abramsky et al., 2001). Nor is this situation only faced by women who have amniocentesis. Any woman who accepts the 'routine' anomaly scan at 18–20 weeks – as virtually all do – may be told of structural anomalies ranging from the severe to the relatively minor. These ethical issues are therefore with us now, not just in hypothetical future scenarios.

Currently, prenatal screening tests are discussed with women at booking. Information is given by midwives and via leaflets. Midwives although key health professionals for women during pregnancy, are busy with multiple complex demands on their time. They know a lot about the care of mothers and babies, but they are not necessarily experts on Down's syndrome, or on the range of conditions that ultrasound scanning points to, nor should we expect them to be. There are ways to inform women and their partners that can be implemented prior to pregnancy and outside the prenatal clinic. There is, in the case of Downs for example, a nationally available booklet now being given to all pregnant women in the UK which contains substantially more information about the syndrome than has previously been the case. This booklet could, for example, be made available in GP surgeries and Family Planning clinics and be part of discussions on sexual and reproductive health more generally. There are some excellent sources of authoritative and experiential information on the Internet that could be more widely publicised (www.dipex.org/antenatal.screening, antenatal screening web resource ANSWER at www.antenataltesting.info) and the opportunity exists to develop some interactive resources.

There is room for information and space for debate within schools, the media and wider social networks in addition to health care settings prior to pregnancy. Health professionals do not have a monopoly of information and a hurried clinical encounter when they are simultaneously explaining procedures and 'requesting' 'informed consent' for screening tests is not the optimum setting for reflection. The rapid technological development of point of care screening and testing leaves less time for informed decision making in the clinical.

The two studies reported here make it clear that a woman does not appear at the clinic as a *tabula rasa*; she usually already has knowledge and beliefs about Down's syndrome and other disabilities and a value system within which this knowledge can be located. We have seen in Study 1 that many women make the decision to accept testing before pregnancy and that in Study 2 they show considerable discrimination in their attitudes with Down's syndrome not being in the 'top four' conditions for which most women would want testing.

The data show that women participating in and agreeing to screening do not necessarily subsequently agree to further diagnostic testing or termination. They do not consider all anomalies in the same way and many have a nuanced understanding of these complex issues. It is important therefore that information is provided outside of the antenatal care provision as well as within it not only about the tests that are available but also about the meaning of these conditions in terms of their impact on the day to day lives of both the children and their families so that women and their partners are equipped to respond to the choices they will have to make.

Innovative screening technologies for many illnesses and conditions are proliferating with increasing possibilities of these being deployed at the point of care quite routinely. This raises two key policy and ethical issues: first, routinisation and the speed of results makes it more difficult for women to opt out of the screening process; secondly, there is a need to develop new competencies in understanding and choosing what to know as well as how to navigate through the troubled waters of prenatal screening.

Acknowledgements

This chapter is based on research supported by two ESRC grants within the ESRC/MRC Innovative Health Technologies Programme Awards: L218252042 on the Social Implications of One Stop First Trimester Antenatal Screening and L218252013 on Social and Ethnic Differences in Attitudes and Consent to Prenatal Testing. Apart from the authors to this chapter, the other members of the research teams were Kevin Spencer, Bob Heyman, Clare Williams, Laura Pitson, Rachel Grellier, Maria Tsouroufli, Howard Cuckle, Robert Mueller and James Thornton. The researchers also want to thank all the women and health professionals who took part in both studies.

3
Genetic Ambivalence: Expertise, Uncertainty and Communication in the Context of New Genetic Technologies

Anne Kerr and Sarah Franklin

Introduction

Recent developments in genetics have provoked considerable controversy about the patenting of DNA, the confidentiality of genetic data and the provision of genetic tests directly to consumers. A range of patients, professionals, industries, policy makers and publics are increasingly closely involved in negotiations about these issues, engendering what some commentators have described as 'genetic citizenship' (Ginsburg and Rapp 2001, 2002). The rights and responsibilities of donors, researchers, and service providers in relation to biobanking, susceptibility testing, preimplantation genetic diagnosis, and genetic patenting raise issues that are described by sociologist Nikolas Rose as 'the politics of life itself' (2001). The complexity of expert systems, the rise of patient activist groups, and the commercialisation of the public sector are part of this story, but the endemic risks and uncertainties of new technological developments are also relevant (Lock and Farquhar 2005; Rapp, Heath and Taussig 2001; Rapp 2003).

The problems of 'genomic governmentality' – for example how to involve more and different constituencies in the development, regulation, and application of new genetic technologies – rely crucially on processes of communication in order to achieve maximum inclusiveness and democratisation. At the same time, many people, including experienced 'genetic professionals' in fields such as genetic screening, do not feel confident 'taking a stance' in relation to debates over the use of genetic technology, the devotion of public resources to biobanking, or personal decisions, such as how much genetic information a couple might want to acquire about their potential offspring in the midst of an ongoing pregnancy. This is not surprising, since many of the situations connected to new genetic technologies are unfamiliar and may be daunting, or simply awkward. However, while much public debate and media coverage of the new genetics is organised in terms of either-or positions, for-and-against debates, or starkly polarised clichés – such as 'the reality vs. the hype' – data from an increasing number of studies of social encounters with new genetic knowledges and technologies reveals that few people, be they users or providers of genetic information, experience its challenges and demands in such 'either-or' terms. Indeed, data from studies from many countries illustrates that ambivalence and uncertainty comprise a significant dimension of many people's relationships to genetic information and technology. While this is by no means the only pattern, it is a significant one that we discuss with reference to several recent studies, and our own research, below.

'Ambivalence' is traditionally defined as conflicting or opposing impulses that are simultaneously present. In this sense, ambivalence conveys the sense not only of mixed emotions, but of *opposite and coincident* views or feelings. A second, equally common, definition of ambivalence is confusion, uncertainty, or hesitation deriving from lack of a clear direction. Hence, the adjectives 'ambivalent' and 'ambiguous' are defined in terms of being unclear or uncertain. 'Ambi' is the prefix denoting 'both' as in 'ambidextrous' – capable of using both hands. Ambivalence is thus similar to equivocation in denoting the presence of more than one opinion or feeling. Significantly, this lack of singularity is often perceived negatively, and on the whole it could be said that ambivalence has a negative connotation. To be ambivalent, and have 'too many' views, is to be deficient in the powers of discernment, judgement, or knowledge capable of resolving them into the 'right' one.

Consequently, synonyms for ambivalence include inconsistency, uncertainty, hesitation and thus unreliability. In terms of expertise, authority, knowledge, judgement, honesty, professionalism or reliability, ambivalence, like ambiguity, would seem to be an anathema. Synonyms for 'ambidextrous' include 'false-hearted', 'duplicitous', 'double-dealing', 'perfidious', and 'crafty', much as 'equivocation' can be seen as 'talking out of both sides of their mouths', being 'two-faced', or not knowing one's own mind.

However, much recent analysis of data collected in the midst of what can be described as the socialisation of DNA – meaning its entry into many more technical and non-technical dimensions of our lives, from clinical and forensic uses to Hollywood film and car advertising – is that ambivalence has a number of positive connotations, as well as negative ones. Thus, in contrast to the view that the 'best' and 'most correct' relationship to genetic information is simply one of clarity, accurate understanding, and rational decision making, ambivalence is in some situations a desired goal of communicative outcomes concerning genetic information, for example when full disclosure might be destructive or unwanted. Similarly, the connotations of indecisiveness, inadequate comprehension, or lack of direction that may be attributed to 'ambivalent' feelings or views are countered in many studies that conclude with calls for inclusion of a greater diversity of voices and positions to maximise the range of different, often conflicting views, in genetic decision-making contexts such as the formulation of ethical protocols for biobanking.

These 'positive' correlations with ambivalence in the context of the new genetics – whereby ambivalent responses are seen less in terms of being inadequate, immature, and irrational and more in terms of being wise, inclusive, and empathetic – raise many questions that will be the subject of future studies. In this chapter we begin this process by exploring this question on the basis of a number of empirical studies by social scientists, as well as our own research.

The first section focuses on Kerr, Cunningham-Burley and Tutton's research into accounts of the political and economic backdrop to the generation of genetic knowledge and understanding, from a range of contrasting constituencies. In this section, the aim is to consider how participants framed themselves and others when expressing ambivalence about what we have called the politics of genetic information. We also explore the different ways in which they sought to resolve and/or sustain ambivalence about these issues. This allows us to reflect upon the nature of contemporary ambivalence about genetics, particularly its implications for further research in this area. In the second section, we focus in particular upon ambivalence about genetic information in clinical and familial contexts, drawing on evidence from Franklin and Robert's study of preimplantation genetic diagnosis (PGD).

The politics of genetic information

For the 'Transformations in Genetic Subjecthood' project, nineteen focus groups were conducted, with between two and five people in each, and several public events concerning genetics were organised. Here, however, we concentrate on data from only twelve of the focus groups, which we broadly classify as 'professionals' (6 groups) and 'lay people' (6 groups) (although as we continue we will query this distinction). 'Professional' focus group members included genetic counsellors, a team of health care practitioners involved with medical genetic services, scientists involved with UK biobank, actuaries with an interest in genetics, university staff with an interest in the commercialisation of bioscience, and representatives of a pharmaceutical company. 'Lay' groups included representatives of support groups for people affected by cancer, genetic disease and Alzheimer's as well as school students, members of the Society of Friends (Quakers) and refugees. We asked a range of questions about genetic testing, patenting, biobanking and public consultation. Although the focus groups are not strictly comparable because they varied in terms of the numbers of participants, range of questions, time for discussion and location, the data gathered from these focus group sessions nevertheless provide the basis for important insights into the ways in which different groups expressed and handled ambivalence about genetics.

Professional groups

Broadly speaking, we found that participants in the professional groups explored their views with confidence, often using detailed examples of their experience. Even when they expressed a lack of 'expert' or technical knowledge about the topic under discussion, their accounts were rich, sophisticated, personal and expressed with confidence.

Some groups and individuals within groups with special interest in genetics (e.g. genetic counsellors) were, as we might expect, much more certain of their views than others. On these occasions, participants gave detailed accounts which included knowledge of the technical possibilities of new research and authoritative views on how it ought to be established, managed and regulated. This tendency was especially evident in the group of scientists with close involvement with Biobank, but it also happened in the group of university staff to a lesser extent.

In all but one case, professional focus group meetings took place at people's workplaces, which may have limited the potential for participants to express ambivalence freely. However, professional groups did express ambivalence about the topics we discussed on a number of fronts. Ambivalence was expressed by some about patenting genetic sequences or discoveries, as opposed to inventions, data protection mechanisms in large-scale research projects like Biobank and 'high tech' solutions to illness, for example predictive testing without targeted treatments. Although some groups suggested that it would be possible to mediate these concerns through schemes such as compulsory licensing, or broader research agendas where genetics was set in context, this kind of ambivalence was frequently unresolved.

There was also considerable ambivalence about participation in the research activities of the Biobank in the professional groups. Participants spoke of their sense of the scientific merit of the study, but also pointed out that they would not want to participate as individuals. For example, one participant in the group of university scientists noted:

> ... it's a difficult one for me in some ways, because I can see some circumstances in which in actual fact I'd rather do the ostrich and bury my head. ... Curiosity driven [research] is fine at the third person level when it comes down to you personally, it's ... [different]

In part this was explained by lack of time, or cynicism about data protection, but it was also linked to several expressions of concern about a lack of personal feedback to participants. This seemed to signal a deliberate split in people's accounts of self, between their professional and their lay selves. However, on other occasions it was related to an alignment between professional and lay selves, for example some participants' positioning as clinicians, whose purpose is to identify and treat disease, meant that they queried large scale research such as Biobank, whilst also expressing a disinclination to participate as research subjects.

On the political front, some professional groups also expressed considerable ambivalence about public consultation, querying the dangers of institutional capture of lay participants and consultation for consultation's sake, and the difficulties of regulating given the pace of change and global reach of scientific research. Other groups were more certain that these problems could be mediated by better public education and clearer regulatory frameworks. This related to their personal sense of 'fair play' as well as their professional experience, as suggested in the following excerpt from the group involved in the Biobank who were discussing concerns that the project did not offer personalised feedback:

> I mean my personal feeling about this ... it may not be same as everybody else's, but, you know, it's like ... anything that you enter into, you know. If you, as long as you are informed properly of the rules ... the terms of engagement at the start of the project, well that; I'm happy with that. So if it was me ... and I'm of the age group ... I will be a participant within Biobank, ... as long as I've been informed, and told that no information will come back to me, I'm happy to do that. I mean if they find that I am carrying some particular ... gene variant, that means I'm going to drop down dead at seventy, so be it. You know, that's the way it is.

These professional groups also demonstrated interesting instances of reflexivity about their ambivalence, or lack thereof. The genetic counselling group, who were all female and did not work together in the same unit, expressed more ambivalence than the other groups, but also commented that they would find it difficult to be so frank about their doubts in a workplace context. This contrasted with the group involved in clinical genetics, whose participants were less open to exploring ambivalence, perhaps because the interview took place in their workplace setting. Other professionals resisted exploring their ambivalence by stating that they had little knowledge of other specialist areas, so they were unable to comment (although they sometimes went on to discuss these issues anyway). A sense of disenfranchisement was expressed by others – and sometimes deliberately adopted – for example the pharmaceutical company scientists said that they steered clear of engagement with regulatory decision making and just followed the rules. Others were more concerned that they had not been consulted about Biobank, despite their expertise.

Lay groups

The lay groups were conducted at a number of settings, including the community venue in which the group usually met, and a range of other meeting places, including the University of York. Most of the participants knew each other before the focus group, but some were less familiar with the other participants (the genetic disease support group members, who represented different groups and the refugee volunteers). These groups engaged with the questions in a sophisticated and informed manner, although they obviously drew on a range of different examples of their experiences as patients and members of the public in order to inform their analysis. This did not, however, mean that expertise was not evident. The group involved in assessing research concerning Alzheimer's disease, was clear about their expertise in the assessment of genetic research, whilst members of other groups, such as the Society of Friends, drew on their scientific and other related expertise in the course of the discussion.

The lay groups were ambivalent about the technicalities of genetic research and services, as were professionals. They expressed concerns about data protection, patenting discoveries and the value of large scale research projects like Biobank in the context of public health initiatives. The position of expert patient/carer also invoked a strong critique of our study in the group of people involved with the Alzheimer's Society, for a lack of clarity about its policy relevance in particular. However, more questions were raised about the predictive value of genetic testing, the models of causation on which they are based, and the need for high tech tests and targeted treatments, than in the professional groups. This occurred from the perspective of potential participants in genetic research and testing, but also from the perspective of an expert patient/carer (support group member), a moral/religious agent, a different kind of expert (psychoanalyst), and a worker with experience of workplace hazards.

As we might expect, there was considerable moral ambivalence in the lay groups as well. One group discussed whether disease was a test from God (refugees) whilst others raised questions about people's responsibility for crime and the influence of genetics (Friends) – these issues were not discussed in the professional groups. However, other issues about balancing altruism and self interest in participation in research projects like Biobank were discussed in both groups. Interestingly, the public groups seemed more willing to participate in this kind of research than professionals. People in several groups signalled their willingness to participate when the issue was first raised. Perhaps one participant's rhetorical question 'don't you think we have to trust?' in reference to sponsoring bodies, like the Wellcome Trust, helps to explain their greater openness. This did not exclude expressions of cynicism and alienation about the political process, however. Once again this was also expressed in some of the professional groups, but it was most apparent in the lay groups, especially in the cancer support group. The school group were also sceptical about public involvement in this kind of genetic research, because we live in a fractured society. As one participant commented, 'Maybe if it was a … closer society, if you were closer with your doctors …' you would be more interested. The lay groups raised similar concerns to those expressed by professionals, although they were sometimes more blunt, especially when questioning the political economy of genetic research and its importance to researchers' careers, rather than the public as a whole. The dangers of surveillance and misuse of data were frequently raised, and seemed to be considered to be inevitable by some, although others suggested similar regulatory mechanisms to those proposed by some professional groups.

The lay groups did seem to be more ambivalent than the professional groups in the sense that they were less likely to express definitive solutions to ambivalence such as standardisation, regulation or individual choice. They also raised more questions about the technicalities of genetic research and tests, and discussed moral ambivalence more fully. However, we also found that they were more willing to contemplate participation in genetic research as research subjects. Although they shared many of the professionals' concerns about the political process, particularly around regulation of patenting and public consultation, some lay groups were more cynical about the possibilities for positive change, largely because of a sense of disenfranchisement. Others were more resigned to trusting experts, although not without some discomfort. These views were expressed from a broad range of positions, not just lay person, but carer, patient, expert, Muslim and Christian. We also found some striking parallels between some of the views expressed in the lay and professional groups. For example, the Biobank scientists and the Alzheimer's disease group both queried the relevance of our research within a broader discussion where they were explicit about what constituted good research practices.

It is interesting that professionals in these groups expressed considerable ambivalence about the Biobank, and that they were far from unanimous in their support of genetic research. Their hostility to patenting is also of interest, although perhaps less surprising given that the majority worked for the public sector, and shared patients' concerns about access to services and treatments. Professionals also seemed, for the most part, not to subscribe to the classic 'deficit model' of public understanding of science, although they were ambivalent about the meaning and nature of public participation in decision making about genetics. Lay people's sophisticated engagement with the technicalities of genetics was also of interest, as were the range of subject positions from which their accounts emerged.

This seems to suggest that the divide between lay people and experts in our study was especially narrow where their ambivalence about the political economy of genetics is concerned. The questions they raised about the technicalities of genetic research and regulation were also remarkably similar. On the other hand, there appeared to be a greater gulf between professionals and lay people in their articulation of moral ambiguity, particularly where religiosity was concerned. Professionals also seemed to be able to marshal more interpretive resources to resolve or discount ambivalence in the course of their discussions. For example, where ambivalence was expressed about Biobank this was often framed as a personal rather than a professional perspective. However, professionals' management of ambivalence did not necessarily mean that they abrogated responsibility in the process, indeed several instances of professionals moving between their public and private selves suggest endemic tensions around their rights and responsibilities in this regard. Lay people appeared to be less able to resolve ambivalence, yet more able to explore it through the adoption of a wider range of subject positions than their professional counterparts. In other respects lay accounts bolstered rather than undermined professional authority. Their faith in regulation, although far from unambiguous, was also clearly evident.

Interpreting genetic knowledge

Authoritative knowledge, judgement, and above all truth are often thought of *as singular*, particularly when scientific 'facts' are concerned. The entire basis of the experimental method on which modern science is based, for example, relies on the ability to determine outcomes based on observable facts, and to analyse these through a process of cumulative induction. Thus, in many of the scientific contexts in which new genetic technologies are being employed it is of the utmost importance to 'get it exactly right'. The presence or not of a gene, or of certain mutations, or even the exact number of repeats within a mutation, may be crucial in determining the likelihood – or certainty – of one outcome or another. The effort to be absolutely sure that when a gene appears to be present it is in fact there is the driving force behind the enormous care taken in contexts such as prenatal screening, or PGD, to ensure there is no contamination, no errors of amplification, and no misreadings, or mislabelings, or miscountings that might lead to a misdiagnosis.

At the same time, it remains the case that, even within science, doubt, equivocation and ambivalence are crucial. The sociology of scientific knowledge has comprehensively shown that processes of proof, replication and experimental design are profoundly social, and, as such open to flexibility and negotiation (Jasanoff et al. 1995). This ambivalence of science is far from unrecognised within the scientific community itself, the hype about discoveries and breakthroughs notwithstanding. It is often argued that the hallmark of a successful scientist is to continually experience intense doubt. Likewise, certainty is seen as one of the worst traits with which to begin a scientific career. Scientific certainties, like French burgundies and Chinese eggs, increase in value as they age.

It is consequently unhelpful to imagine that we can distinguish between something like the social or subjective benefits of ambivalence, for example in the face of the kinds of impossible choices many would-be parents have to make about a wanted but chromosomally abnormal pregnancy, and the context of scientific authority, where ambivalence is only ever compromising.

Indeed, it is a possibility suggested by a number of recent studies that the reverse is true.

For example, in her path-breaking monograph based on over 15 years researching women's experiences of amniocentesis in New York City, anthropologist Rayna Rapp (1999) found that ambivalence and ambiguity came in mixed dosages with certainty and singularity of purpose on both sides of the clinician/scientist vs. patient/family member divide. Hence, in some cases where an 'ambiguous' genetic diagnosis was presented, about which little was known, and therefore no professional certainty could be expressed, patients often had strong and unequivocal views – either to terminate a pregnancy, or to retain it. Medical anthropologist Kaja Finkler (2000) made a similar finding in her study of the way in which the 'objective truth' of genetic information enters into the highly emotive and subjective realm of negotiating kin ties. In the realm of decision making about tracing genetic predispositions through genealogical relations, for example in terms of deciding whom to contact for a blood sample, ambivalence was often expressed as a form of care: on the one hand it might be beneficial to encourage other family members to be tested, but it might equally cause them the strain of unnecessary anxiety. This is especially true given the highly ambiguous nature of much genetic susceptibility testing, which may or may not yield reliable data, and, even if it does, may not be correlated to either certainty about the future, or beneficial treatment (Franklin 2003).

These same points are demonstrated in a recently published study by social anthropologist Monica Konrad (2005) on 'the new predictive genetics' in which one of the central themes is the necessity for communication about these new technologies to be composed out of multiple and varied knowledges, voices and expertise. She claims one of the most important aspects of future health care campaigns concerning genetic testing will be the necessity to include users and patients on research drafting boards and policy committees who can provide a variety of different languages of genetic 'facts' for different constituencies. Describing the dilemma of how to allow for people who do not want to know about their genetic 'susceptibilities' or predispositions, while also knowing when to provide this information for others, Konrad notes the importance within genetic counselling of allowing counsellees to remain 'unresolved' – much as this contradicts the rational choice ethos of the entire counselling process (2005: 66–7). Such a model could be described as ethical ambivalence, or professional equivocation, that is in the interest of serving a diverse clientele.

As these studies and the evidence of the previous section highlight, it is less and less accurate to generalise about the process of transmitting genetic information in terms of established models, norms or protocols in which professionals communicate unambiguous genetic risk information to individuals or patients in the name of increased, and/or more rational choice. Different publics versed in their own 'expert' knowledge beliefs must also be involved actively in educating policymakers if 'evidence-based care' is to have any formative influence in the evaluation of emerging initiatives, interventions and technologies.

As Konrad points out, this tendency for health professionals in the context of the new genetics to become highly skilled in shifting registers of speech, language, expertise and tone is repeatedly documented in Rapp's account of the communication 'brokers' in the clinics where she conducted fieldwork. In addition to the need to be able to become, as it were, ambi-literate, in order to 'switch-speak' within and across several knowledge registers, other situations of responsible care and professional obligation towards the disclosure of genetic information require it purposefully be made even *more ambiguous* to patients. In one of the most complex narrations of what might be called the *deliberate promotion of genetic ambivalence* by clinical professionals, Konrad describes the process of 'exclusion testing' for Huntington's disease using pre-implantation genetic diagnosis. In these cases, 'prospective parents who know they are at risk but crucially who *do not want to have their own genetic status made less ambivalent* (i.e. verified as positive or negative) through the selective screening procedures, pre-implantation diagnosis by non-disclosure effectively cuts off the flow of genetic information' (2005: 126).

Thus the field of new genetic knowledges and technologies, and in particular what we might want to call the new explicitness of 'genetic meaning', is replete with examples of ambiguity and ambivalence. This is caused in part by the phenomenon Franklin and Roberts have referred to as the 'genetic gap', meaning the difference between what genetics means in one context (e.g. a laboratory) and another (e.g. a marriage) (Franklin and Roberts 2002). Because, as good genetic counsellors know, these contexts cannot be 'resolved' in order that any one, singular, 'objective' genetic 'truth' prevails, genetic knowledge must always be acknowledged to have multiple meanings, which are equally and simultaneously true, and may well point in divergent directions. This truism also derives from what Konrad describes as the 'fundamental contradictions' (2005: 66) inherent to both the ethics and pragmatics of genetic knowledge transmission, especially of the expert, authoritative kind. For example while such information must be treated as personal and individual, and patients' autonomy protected, this goal is contradicted by the fact of genetic information inherently having meaning for others. Similarly, as Konrad notes, the very existence of tests for mutations such as the Huntington's gene creates a new chronic condition for sufferers, which is the inescapability of choosing whether or not to be tested.

These 'fundamental contradictions' require that ambivalence is sometimes promoted and protected as a component of best clinical practice and conscientious ethical care in the context of clinical genetic communication events, such as susceptibility testing. In these contexts, the use of ambivalence *as a form of protection* acquires a more positive connotation, as does the 'ambi-vocality' of the effective genetic counsellor.

A suggestion of many studies of the new genetics is that much greater diversity of knowledges and expertise must be incorporated into the development and provision of new genetic technologies, especially in terms of health care and reproductive decision making. This tendency, in a sense to exploit the 'genetic gaps' to achieve maximum coverage, also needs to occur as far 'up the pipeline' as possible in order to be effective – for example by involving patients in the design of scientific research, as has already begun to take place in some single-gene disorder patient groups (Rapp, 2003). It is in this way that 'promoting ambivalence' may be linked to democratisation, or the 'biological citizenship' described by Novas and Rose (2000).

Yet another dimension of ambivalence derived from Franklin and Robert's study of genetic knowledge in the context of pre-implantation genetic diagnosis, based in the Assisted Conception Unit of Guy's and St Thomas' Hospitals (where part of Konrad's study was also based) was a somewhat unexpected correlation between uncertainty and trust. This finding was described as 'uncertainty value', denoting PGD patients' appreciation of clinicians' willingness not only to discuss, but to emphasise, the many uncertainties involved in PGD.

In her fieldnotes from her first visit to the PGD clinic at the Leeds General Infirmary to begin ethnographic fieldwork in 2001, Sarah Franklin described her surprise at the explicit attitude of clinicians, nurses and genetic counsellors that their duty was to present 'the really bleak picture' of PGD, thereby 'putting the fear of God into them':

> On my first afternoon, [the PGD nurse coordinator] took me to several meetings between genetic counsellors and prospective PGD patients. I am amazed how much failure is emphasised, how much they go on about how many things can go wrong, how it is all very early days for the technique and for the clinic, and how complicated all of the procedures are. They always mention an earlier misdiagnosis, in one of the clinic's two attempts so far, producing what they refer to as a '50% possibility of misdiagnosis'. They seem to want patients to be almost adamant they'll do nothing but PGD. Later a couple comes in for their third consultation. They are incredibly articulate and clear about their reasons for wanting PGD. The genetic counsellor says to them: 'This technique is for people like you – people who know PGD is not the easy option, people who know this is a very difficult technique.' (09/01/01)

As Franklin and Robert's study showed, willingness to discuss the drawbacks of PGD – a dauntingly technically complex procedure requiring both microsurgery to remove a single cell from the 8-cell embryo and successful amplification of its minute amounts of DNA – correlated for many PGD patients with increasing confidence in the PGD team. As Ben and Sally, one of the couples at Guy's and St Thomas' Centre for PGD who were interviewed for the study, described their experience:

> Ben: And so we had a talk with the [consultant] and the genetic counsellor from the [PGD] clinic ... They [explained] how they do it, this attaching probes to the certain genes and then looking to see this, that and the other, and then they talked us through all the, um, additional tests we have to have afterwards because it's, um, my particular translocation's biggest setback is going to be Down's. So [the consultant] said obviously this isn't 100 percent guaranteed that you won't have Down's because it's only a 2 dimensional picture that they take. So, there could be genes lying underneath other genes that they don't see ... So, he talked us through all these various ... amniocentesis, CVS tests and this, that and the other, and the ... *additional* risks that they bring into it, and would we be willing to ... have the test done?
>
> Sally: It was a lot clearer that time, wasn't it, to understand? [...] I suppose it was like, you think he's just going to take [your eggs], sort your embryos out, and give you children. I think he wanted to make sure that we understood that things could still go wrong. It was very early days and um, it, it was helpful wasn't it? It really was! 'Cause we came out of there feeling quite refreshed actually 'cause we actually both felt we understood. Having spoken to the [consultant], and the genetic counsellor, we both understood fully what it meant.

As Ben and Sally explain, understanding everything that can go wrong, and having one's expectations knocked back at the outset of treatment is described as 'helpful' and 'refreshing'. Emphasis on the number of different technical obstacles which may prevent successful treatment, or lead to a misdiagnosis, and the additional risks introduced by prenatal testing are described as making the technique 'a lot clearer [to] understand'.

These and other examples indicated that, at least in the context of PGD, patients valued clinicians' candid appraisal of the number of 'unknowns' involved in treatment particularly insofar as it allowed them *to manage their own uncertainty rather than having decisions made for them*. The effort by clinicians and genetic counsellors to foreground the uncertainties of treatment *increased patients' trust in them* and correlated directly to the high esteem for their professionalism. In some cases, the ability of members of the PGD team to convey the risks and uncertainty surrounding aspects of the technique, such as the unknown causes of implantation failure, led to patients describing them as 'not really like doctors at all'. In a statement that neatly captured the normative incommensurability of uncertainty, or 'ambivalence', and authoritative medical expertise, clinical best practice and professionalism these cases clearly contradicted, Tony and Melissa, another couple from the

Guy's and St Thomas' Centre for PGD, expressed their appreciation of the team through a series of double, positive and 'backwards' negatives that denote praise.

> Tony: I don't look at them as doctors. Do you?
> Melissa: No.
> Tony: ... Not 'cause they're not professional or nothing, just because they're really, really good to you. They never ever speak down to you ... And they involve you with everything, which is great for me!
> Melissa: They do explain everything and if it hasn't been successful then when we go back there the file's out, they go through everything – what they think went wrong and what's this. They're all really good.

As Tony and Melissa emphasise, a positive sense of involvement is produced by thorough and lengthy explanations of failure. The experience of not being spoken 'down to' prompts a comment that the PGD team do not seem like doctors at all. The form of this comment, as a question, elicits an affirmative negative: 'No' (meaning 'yes, I agree, they don't'). This is quickly followed by a positive qualification: 'it's not because they're not professional' – a double negative that is followed by 'or nothing'. These 'knowing' comments dialogically invoke the stereotypical patronising medical authority figure, and the polarised view of the medical expert (who has knowledge) and the ignorant patient (who has none) as the background to the comments that 'they're really good to you', 'they never ever speak down to you', and 'they involve you with everything'. Rather than being presented with a single, resolved, decisive version of 'the path ahead' through what one PGD patient described as 'the topsy-turvy world of PGD', the value of more uncertain, equivocal and *ambivalent* accounts of its high potential failure rates and numerous 'unknowns' was both increased trust and a more confident understanding PGD. For PGD patients, clinicians and genetic counsellors alike, the value of these *ambivalent accounts* lay in the fact of their being seen as more responsible – and indeed as professional best practice in the midst of often ambiguous genetic information.

Conclusion

As the many examples we have discussed in this chapter illustrate, there is a degree of evidence that 'ambivalent' responses to the new genetics comprise a significant social phenomenon, which may reflect particularities of the genetic 'turn' that will be better characterised by future study. While it is possible to demonstrate also that these 'ambivalent' responses can be correlated to positive 'goods', such as better communication, increased inclusiveness or greater trust, the two studies by Kerr, Cunningham-Burley and Tutton and by Franklin and Roberts, also demonstrate that ambivalence may take a number of patterns and serve a range of functions, which, perhaps fittingly, may not resolve into a predictable pattern. Indeed, these two studies usefully depict a range of responses that suggests while the 'ambivalence patterns' characteristic of new contexts of genetic knowledge and technology may indicate a prominent form of interaction or positionality, this in and of itself may not show a consistent correlation with specific meanings, values or 'feelings' about the new genetics. Crucially, social actors' constructions of, and responses to, genetic information involve the sophisticated expression and management of ambivalence according to the complex relationships and contexts in which they are located. At times, the expression of ambivalence opens up space for reflection on the ethics of care within the family, or professional responsibilities to patients, or even one's obligations and duties as a citizen. At other times accounts of ambivalence signal frustration and alienation from biology and/or politics, which cannot be resolved. It is not the case that lay people are the victims of ambivalence and professionals are its nemesis. Ambivalence about the interpretation and politics of genetic information can be suppressed, discounted, managed or foregrounded for a variety of professional and personal purposes, across a range of clinical, political and social domains.

As a component of the emergent 'biosociality' first described by anthropologist Paul Rabinow (1992) in reference to the ways in which new forms of genetic information will become inculcated in, and be reshaped by, emergent social organisations and affiliations, such as patient groups, 'genetic ambivalence' remains a distinct, and in many ways prominent finding to emerge out of the first generation of 'socio-genetic studies' using empirical methods to investigate actual genetic encounters and the wider politics of genetic information.

Acknowledgements

The authors would like to acknowledge the contributions in regard to data collection and analysis of co-researchers Sarah Cunningham-Burley, Richard Tutton and Celia Roberts.

Part 2
Information and Empowerment

4
'Pathways to the Doctor' in the Information Age: the Role of ICTs in Contemporary Lay Referral Systems

Sarah Nettleton and Gerard Hanlon

Introduction

Just over three decades ago Zola (1973) published a paper with the title 'Pathways to the Doctor: from person to patient'. This seminal work became one of the most influential and cited articles in medical sociology. Based on an empirical analysis of patients' accounts of their reasons for attending a hospital outpatients in the USA Zola identified his five now famous 'triggers' (see below), which prompt consultation with formal health care providers. But of course there have been substantial socio-economic, political and technological transformations since Zola carried out his fieldwork. The late 1960s and early 1970s was a period that is characterised as being an era of Fordism, modernism, professionalism, industrialism and so on. This is in contrast to the post-Fordist, late modern, consumerist, information age which is presumed to more accurately capture the features of contemporary life. The aim of this chapter is to reconsider the work of the likes of Zola and his contemporaries in order to assess whether and to what extent his analyses of 'pathways to the doctor' still have any analytic purchase in relation to people's routine experiences of health care. The chapter argues that while continuities can be discerned over the last 30 years there are also changes; imperceptible perhaps at the level of everyday practice but occurring nevertheless. One of the more perceptible changes is the fact that 'patients' are more likely to be seeking health care advice from nurses or the Internet rather than the doctor, and in domestic or virtual locations rather than hospital settings. In this chapter we present findings from two empirical studies which examined people's experiences of these health resources. We begin however by placing our discussion in context by drawing on conceptualisations of health care provision.

Help seeking, health care and modernism

In the post-war years social scientists and health policy analysts identified an unanticipated feature of help-seeking in relation to health care. And that was that people who are ill, diseased or sick do not invariably seek medical help. This phenomenon was encapsulated by the concept of the *'clinical iceberg'* (Last, 1963; Wadsworth et al., 1971). This prompted sociologists to explore how and why individuals do, or do not seek medical help. Zola's study of the reasons given by outpatients at a New York hospital is one of the most well known of these explorations. He revealed that it was not symptoms per se that prompted people to seek help but rather it was their social circumstances. He identified five 'distinct non-physiological *triggers* to the decision to seek medical aid'. These were: the occurrence of an interpersonal crisis; the perceived interference of an illness with social relations; 'sanctioning' by another person that a visit is warranted; perceived interference with physical activities; and temporalizing of symptoms – 'if its no better by Monday'. The patient of the 1950s portrayed within the Parsonsian (1951) *sick role* thus became more active and socially embedded.

A sick person it seems does not simply seek help but manages his or her symptoms and interprets them within their wider social milieu. Furthermore, 'going to see the doctor' as Stimpson and Webb (1975) revealed in their classic study was a complex social process whereby patients planned, rehearsed and negotiated their approaches to their subsequent interaction with professionals. People sought advice from non-professionals; from friends, family, magazines and so on. In other words medical and health advice was proffered and sought within what came known as the *'lay referral system'* (Freidson, 1970). Sociologists therefore cast light on the informal health care work undertaken by lay people and demonstrated that in seeking formal health care professionals saw only the tip of the iceberg of illness (Stacey, 1988).

Zola's triggers, the sick role and the lay referral system are then concepts of the modern age. This is an age that Pickstone (2000) conceptualises as *productionist medicine* and *communitarian medicine*. Within this context the dominant organisational forms of health and medical care were characterised by professionalism, public service and welfare and were, in turn, orientated towards protecting and maximising the populations' health. Motivating, and so understanding why patients did or did not see their doctors is important in this context. Arguably, today however we are in a different era. Pickstone for example suggests that *consumerist medicine* has come to predominate with people and patients becoming more demanding and discerning and expecting a wide range of services, treatments and so on. Policy makers expect and are indeed keen to capitalise on patients; who are 'a great untapped resource for the NHS' (Nowlan 2003). Health care provision has become more diverse and fragmented. For example, complementary and alternative medicines vie with traditional biomedicine; health care can be sought in 24-hour 'walk in' health centres, at the end of the telephone or by logging on to the Internet. The patient too has become a multifaceted figure; informed patients, 'expert' patients, health care users, health consumers, reflexive health seekers all readily acknowledged to have their own legitimate demands. Last's (1963) clinical iceberg has become inverted and the 'problem' for health policy analysts or the 'question' for researchers is how do we 'encourage', or how do we 'understand' the management of health and illness by lay people.

Conceptualising health care in the information age

Echoing Pickstone's periodisations of medicine, Smith (2002) (a medical doctor and former editor of the *British Medical Journal*) conceptualises these changes as a move from what he calls 'industrial age medicine' to 'information age health care'. As we can see from Figure 4.1 there is something of an inversion of the clinical iceberg. In the context of industrial age medicine people were encouraged to seek professional help with the management of their health and illness, by contrast within the context of the information age self care is privileged. Such changes have reverberations for health care

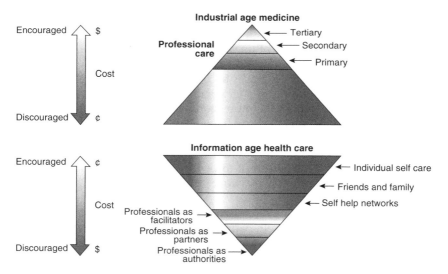

Figure 4.1: The future of medical education: Speculation and possible implications

Source: This figure is included in the presentation by Smith (2002) (15.7.2002) www.bmj.com/talks) but is an adaptation of a figure in Health Canada – Health Care Network (2003) *Supporting Health Care: the contribution of nurses and physicians* http://www.hc-sc.gc.ca/hppb/healthcare/pubs/selfcare/index.html

and medical practice. As Horton (2003) the editor of the *Lancet* observed:

> [T]he most fundamental change between past and present medicine is access to information. There used to be a steep inequality between doctor and patient. No longer. As people understand the risks as well as the benefits of modern medicine, we increasingly desire more information before we are willing to rely on trust to see us through. This need to be transparent about what doctors know (and what they do not), to engage in a consultation on closer to equal terms with patients, has changed the way medicine is practised. (2003: 40)

Doctors have to recognise that their patients may be 'smarter' than they are; they have to work in networked organisations rather than hierarchical ones; they must acknowledge that their clinical knowledge can be patchy; and they must make use of information tools and systematic reviews of evidence rather than just relying on their clinical experience. These views echo that of Blumenthal (2002: 526), a North American analyst and physician, who writes:

> A decade ago it would have been unimaginable to suggest that the medical profession might be headed, if not for extinction, at least toward a profoundly diminished role and status in ministering to society's ills. Yet the information revolution, coupled with other recent developments like the rise of alternative types of health care personnel and the new health care consumerism, has made such changes seem not only imaginable but even a plausible extension of prevailing trends.

So just as the figure of the patient is becoming more varied so the figure of the doctor is also becoming less clear cut, and medical expertise, which formed such a key feature of the professional within the modernist context, is becoming less secure. Or at least so it would seem, if these theorisations are correct. Interestingly these leading voices within the medical profession cite sociological theorists on globalization, consumerism, the changing nature of expert systems, the proliferation of information and communication technologies (ICTs) and so on, with the work of Giddens, Beck and Castells being particularly foregrounded.

These social theorists describe the metamorphosis of modernism into reflexive modernisation wherein patients, citizens, individuals become more individualistic and reflexive and welfare providers become more flexible, less paternal and more responsive to their needs, demands and wishes. Central to this is the emergence of a new type of individual and a reshaping of the lay–professional worlds. Increasingly, so the story goes, we are less trusting of experts, more individualised and yet somewhat paradoxically more dependent on professionals. Given the dominating influence of the writings of Beck and Giddens it is not surprising that theorizations of engagement with expertise via ICTs have been shaped by notions of reflexive modernization and of 'reflexivity' in particular (Slevin, 2000; Beck-Gernsheim, 2002).

These conceptualisations have led some analysts to suggest that health care users may be empowered by the availability of diverse forms of support (Gillett, 2003; Hardey, 1999, 2001; Loader et al., 2002). Other commentators however have a more sceptical take and are concerned that lay people in particular may become overwhelmed and overburdened by the sheer quantity, and more worryingly, the disparate quality of information and support (Charatan, 2002; Graber et al., 1999). Elsewhere (Nettleton et al., 2005) we have referred to these perspectives as the 'celebratory and empowering' and the 'concerned and dangerous'. There are however other commentators (who it is worth noting draw heavily on empirical data) who advocate a more 'contingent and embedded' take on the use of innovative information technologies (Broom, 2004; Henwood et al., 2003; Savolain and Kari, 2004; Ziebland, 2004; Williams et al., 2003; Nettleton et al., 2004). This position, with which we concur, purports that for the most part, lay people are well able to make 'reasonable' assessments about what constitutes appropriate information when it comes to their own health and illness. It also argues that use of innovative technologies is likely to be shaped by the broader social circumstances in which people live their lives. In relation to health care within the UK, the availability of e-health (re)sources and the development of the nurse-led telephone advice service, NHS Direct, are technologies which are particularly apposite examples of innovative information technologies. These are being supported by a government keen to privilege patient choice, which in turn serves as a justification for the need to provide more flexible and responsive services (Department of Health, 2004). Reflecting the shift from productionist to consumerist and what we might call informational medicine, recent health policy documents highlight the need to shift power from professionals to patients and to ensure that services are tailored to their needs. Examples of consumer orientated health care in the UK include: 24-hour 'Walk in' health centres – designed to meet the needs of the patients not just the professionals; single electronic patient records which will enable patients to move more effectively between the various services on offer and as we have seen the telephone and online services NHS Direct and NHS Online. It is envisaged that by

> 2008, patients will have the right to choose from any provider, as long as they meet clear NHS standards and are able to do so within the national maximum price that the NHS will pay for the treatment that patients need. Each patient will have access to their own personal HealthSpace on the Internet, where they can see their care records and note their individual preferences about their care. (Department of Health, 2004)

Contemporary policies therefore facilitate reflexivity: self-learning, self-management and self-care of health and illness. They encourage and, policy makers would argue, also enable individuals to take responsibility for their own health and health care. Thus they may become empowered so that they can be responsible citizens although from the point of view of resource use this empowerment may be problematic (O'Cathain et al., 2005). There is no doubt that these sources of information are being used extensively, that the majority of the UK population are able to access the Internet and a significant proportion of these use it to seek health related information. In 2004 NHS Direct was predicted to receive almost 14 million calls (National Audit Office, 2002).

Seeking health care in the information age

It seems then that there is something of a convergence between the observations of social theorists and the aspirations of policy makers. Patients and lay people desire and have a right to information and accessible flexible services. And indeed they are engaging with them. But as yet we are only beginning to understand how they engage with these new organisational and technological forms. Are the health seeking activities described by Zola, Stimpson and Webb, Stacey and other sociologists decades ago being transformed? Or perhaps, are they still recognisable today albeit in a modified form? The findings of two empirical investigations, one into users' experiences of NHS Direct and the other into the use of health information on the Internet more generally carried out as part of the IHT Programme, would seem to suggest that the latter is the case (Hanlon et al., 2003; Nettleton et al., 2004). In the light of contemporary preoccupations with reflexivity, choice and consumerism we need to be cautious lest we forget the findings from the classic medical sociological studies. These, as we note above, revealed: how people sought information from multiple professional and non-professional sources (Stacey, 1988); how they actively engaged in 'information work' (Conrad and Strauss, 1985); how they strategically rehearsed their views prior to consultations with health care practitioners (Stimpson and Webb, 1975; Tuckett et al., 1985); how they sometimes did, and yet oftentimes did not, follow the advice and recommendations of the 'experts' (Conrad, 1985). But no doubt then, as today, most lay people still claimed to, and indeed very often feel a need to, 'trust' health professionals and in particular doctors (British Medical Association, 2003). Furthermore as Shilling (2002) points out the cultural underpinnings of the sick role endure; most particularly the value of 'instrumental action' – the idea that something needs to be done in the face of illness.

What we are suggesting therefore is that in the context of contemporary technological, social and cultural transformations there are, perhaps not surprisingly, elements of both continuity and change. In many respects it is the changes that are the most readily visible: the click of the mouse on the 'net' to find out information on virtually any health issue, the ability to secure an informed response from a clinical practitioner 24/7 by simply making a telephone call – very probably on a mobile phone. Clearly availability is socially patterned with access to the Internet in particular mapping straightforwardly onto socio-economic divisions. The availability of more heterogeneous and flexible forms of care however overcomes some of the issues of access which figured prominently 30 years ago. For example, one the major factors which hindered access to emergency and 'out of hours' care for those who were socially disadvantaged was repeatedly found to be vandalised telephone boxes (Holohan, 1977). Today households without a telephone comprise a very small minority and the majority of households now have access to the Internet.

Cognisant of the literature on seeking health care and advice we aim to highlight a number of salient issues which transcend the findings of studies into the use of NHS Direct and online e-health information. Some of the findings may appear banal. We make no apologies for this and indeed an overall theme of our work is just how rooted the use of these technologies becomes in the mundane realities and routines of people's lives. There is little fit between empirical data on experiences of telephone and online support (Nettleton et al., 2004; Hanlon et al., 2003) and some of the claims made by social theorists.

Lash (2002) for example, argues that the emergence of ICTs means that 'knowledge' metamorphoses into information, and becomes 'disembedded'. Engagement with such information is characterized by 'real time relations' whereby the 'reflex' of accessing and reading information becomes fused. On the contrary, it appears that use is actually more contextually specific, it is often contingent upon a perceived need for health information and so the circumstances of use tend to be both *embedded* and *embodied*. Seeking advice and information is *embedded* in people's approach to seeking help, advice and information more generally and the Internet is routinely placed alongside other (re)sources. In this respect it constitutes an extension of the 'information work' (Corbin and Strauss, 1985) which forms part of the routine, everyday management of health and illness and the social processes associated with illness and help-seeking behaviour which we described above. People draw upon a range of resources such as relatives, books, magazines and friends, and it appears that Internet use is becoming thoroughly enmeshed with these processes. Seeking help, we suggest is also *embodied* because very often it is contingent upon specific health needs and so can contribute to the management of care. For example, due to its spatial and temporal flexibility and the desire of young people to talk to someone and be reassured in their new role NHS Direct is used mostly by young parents (Goode et al., 2004a). Information, knowledge and explanations may result in people feeling more reassured about illness, prognosis, treatments and so on, and as such can profoundly impact upon both the emotional and pragmatic aspects of the illness experience, making people feel more in control and so more confident in their response to symptoms (Zeibland, 2004).

Pathways to the professional in the information age

It is axiomatic within the sociology of health and illness that the route to consultation with health practitioners is a social process. Figure 4.2 sets this out diagrammatically. This diagram is typical of any that might be found in an undergraduate text on medical sociology. But it serves to remind us of the context in which innovative organisational forms such as NHS Direct and e-health information technologies reside and may be located from the point of view of lay users. From analyses of people's experiences of their use of the 24-hour telephone call centre NHS Direct and the Internet a number of changes or perhaps more appropriately *modifications* and *extensions* to this social process can be identified. We briefly summarise six below.

(1) First, these technologies may *enhance and extend the lay referral system*. For example, prior to a formal 'traditional' consultation with a general practitioner symptoms may be checked to ensure that they warrant a visit. With the information sourced from the Internet some people reported how they made a provisional diagnosis that could be confirmed by a telephone consultation. For example, one woman who was suffering from intense headaches and sickness was advised by her friends that it might be migraine, and so she went online to seek out further clarification so that she would not, in her words, 'have to waste the doctors' time'. In another household an 11-year-old girl looked up her sister's symptoms on the Internet and diagnosed shingles. The mother rang the GP to confirm the diagnosis thus avoiding the need for a face-to-face consultation. As a mother from the poorest area in the sample pointed out; 'it's here at home and you just click and away you go'

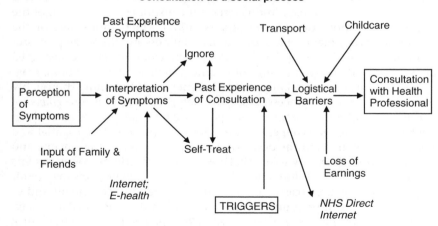

Figure 4.2: Consultation as a social process

which is much more 'handy than making an appointment and traipsing into the doctors' (Interviewee 21).[1]

We can see therefore that e-health information and NHS Direct may be used to facilitate access to health professionals in more conventional settings. For example users of NHS Direct tell us that people are using it to sanction their use of health care more generally or to sanction and empower them to request further clarification. In many ways, this reflects older patterns of health care use wherein people interpreted their own bodies and used others to legitimate a trip to the doctor. Now they are doing this but increasingly feel the need to use another professional to empower them to see a professional that used to be more readily available – the General Practitioner. As a woman in her mid-twenties concerned with her young child said,

> I then called my doctor but the clinic wouldn't, they didn't think it was necessary for me to see anybody because I'd seen a doctor that morning and I said to them 'well he's got worse and the NHS Direct has recommended that I see you' and after a bit of hesitation they then let me see a doctor. (Interviewee 34)[2]

(2) A second modification to this social process is the finding that the *cultural assumptions, norms and values of the sick role prevail*. For example, in line with much of the older literature on patients, callers to NHS Direct were eager to preserve their status as deserving patients. They were very concerned not to be seen to waste the doctors' time, not to use up resources unnecessarily and to disassociate themselves from other 'undeserving' patients (see Goode et al., 2004).

(3) Women have always done the bulk of the health care work within both the public and private domain (Stacey, 1988). Historically male access to health care has been weaker than that of females. Women have also had the role of providing care for others such children or the elderly. Thus women have been socially embedded in health care in ways that men have not (Umberson, 1992). Although computer mediated technologies are male dominated this is not the case when the use is for health related issues. The overwhelming number of callers, nurses and assistants in NHS Direct are female although the service appears at this early stage to be more male friendly than others (Hanlon et al., 2003) and this was also borne out in interviews carried out with parents' use of the Internet for health issues. It seems therefore that a third modification related to use is the extent to which *negotiation of health care remains gendered*. However there are indications of change. For example, one of the ways that NHS Direct may be altering use patterns (Goode et al., 2004a) is by increasing the use of the service amongst men both for themselves and for others. The study of NHS Direct found some albeit tentative evidence that patterns of use may be different for this form of care. Thus of our calls relating to men only 32 per cent of them were made by the

men themselves which of course confirms previous views of gendered health care but men were almost as likely as women to ring about spouses, they made one in three calls about aged parents and one in five about children. In particular, men were most often involved when there was a perceived need to be assertive. The following interview extracts demonstrate how males and females divided up their roles. The first interviewee is a woman who describes how her husband dealt with medical professionals after they felt their young son was not being attended to properly despite his mother's efforts.

> Since we've had the little boy, he's been a lot more assertive in that way ... once James had been ill and he wasn't getting any better, and so he said, 'Right, I'm coming home early and we'll both go in this time'... I'm not as assertive as my husband really. I tend to, 'Oh right, yes, whatever you say doctor.' Of course my husband isn't. He's got a hundred and one questions, and he won't leave the surgery until they've been answered. (C02)

This second interviewee (a male) argues how he was assertive after seeing attempts to dissuade him from getting home health care for his ill partner.

> I simply had one aim at that point, which was to get a doctor out to the house without putting the phone down ... everything was pretty much arranged in the one call. It was acknowledged that things were bad and that a doctor would be calling tonight ... I guess I was being pretty direct, like 'She is sick and she must be seen.'

Hochschild's (1983) *Managed Heart* and others (e.g. Morgan, 1996) suggest that men and women perform a double act wherein women act as ameliorating and when this does not work men come in as assertive or forceful. The experiences of NHS Direct users indicates that men do this increasingly for adult children living at home and for their young children. Thus NHS Direct because of its ease of access and its 24-hour nature may be providing an avenue for men to enter the world of health more than in the past. But before we welcome in a new era, we should bear in mind that much has remained the same in men's health. Men tended to access health care only after they were cajoled into doing so by females in their lives – in ways reminiscent of Zola's calls in study. For example, the following middle-aged man with a wife and a female adult daughter made this typical comment:

> It's a fact that I wouldn't have rang up in the first place if I hadn't been bullied (laughs) – 'you've got to do this, you've got to do that, you've got to take care of your eyes' – stuff like that. I'm one of those: 'Oh, it's all right!' (C29)

(4) As we discussed above, a number of commentators (e.g. Blumenthal, 2002; Horton, 2003) claim that one of the most fundamental changes

between past and present medicine is the greater access to information on behalf of non-professionals that has enabled patients to engage with doctors on more equal terms. From accounts of users' experiences it would appear that this is indeed the case. But again we must be cautious about the extent to which this may be altering interactions with practitioners. For example, our study on the use of the Internet revealed that a major use was to *supplement, and sometimes even enhance their consultations*. The study participants said they would trust doctors more than material they had sourced from the Internet. However, being more informed gave them the confidence to ask the 'right questions' and because they were more familiar with the medical terminology they felt that they could better appreciate the answers. One woman interviewed for this project whose mother died of pneumonia described how she incorporated the online information into her experience:

> I discussed what I'd read [on the Internet] with my sister, and it helped me when I talked to the doctor to understand a bit more what the doctor was saying. It gave me a much better understanding of the prognosis and the likely outcome having read it. More so than I got from the medical staff. I mean eventually they did say 'its not looking good', but I'd already realised that. (Interviewee 11)

We can see how the Internet adds a further dimension to the lay referral system whereby people consult with their friends, relatives and so on.

Following a clinical consultation people would look on the Internet to clarify the advice they had been given, and seek further information on drugs and diagnoses. Again the following example illustrates the integration of information sourced from the Internet into the lay referral network.

> My friend's mum was diagnosed with an aneurism, and we didn't know what it was. I know she wasn't advised what it was, she'd gone to the hospital with her mum when she came home, I said 'Well what is it?' and she said 'I don't really know' and I said 'Well I know it's serious but ...' and we looked it up on the, on the Internet. Because it [the diagnosis] was such a shock to them, that although the consultant had explained it they hadn't really taken it in. (Mother, middle area Interviewee 24)

On some occasions people simply felt that they had not been given enough information and so needed to supplement what they had been told in the consultation. For example, a woman who was informed by her GP that she had a stomach ulcer;

> I wasn't really told a lot about what I had, I was just sort of given medication. I wanted to look into it a little further, just so that I could understand what it was. (Woman, wealthy area, Interviewee 7)

We know from existing research that people seek clarification about treatment regimens and often experiment with medication (Conrad, 1985). Thus the availability of online information is not the reason why people seek to question, clarify or reassess proscribed and prescribed treatments but rather acts as a resource which facilitates such action.

(5) Another dimension of assessing information is the way in which people *strategically 'manage' their information-seeking and the distribution of the information they source*. For example, they would limit the amount of information they sought about a condition especially if they were aware of long-term complications. One mother, who sourced information about diabetes for herself and her son, was careful to avoid seeking too much information on prognosis and possible complications, preferring instead to concentrate on issues relating to the current management of the condition. Another user who sought information for a friend whose son had been diagnosed with cancer explained how she read the information through first and passed on only what she considered to be appropriate in the light of her own husband's recent experience of the condition.

(6) The ready availability of information is, as we noted, a source of concern amongst some health practitioners and commentators. Patients may become misinformed, duped or misled. However what is striking from our studies is the extent to which participants are keen *to trust and depend on their doctors and adopt strategies to seek out 'trustworthy' information*. Within the interviews undertaken for the study of Internet-use all the study participants expressed concerns about the quality of information and the dangers of misinformation. Their views echo the anxieties articulated within the medical literature. People were aware that: inappropriate treatments could be fatal; commercial companies may dupe vulnerable users; websites and other information sources may be created by quacks masquerading as medical practitioners; and so on. As one mother put it;

> I think too much knowledge can be dangerous. It makes people think they can treat themselves, rather than getting medication from reputable sources. (Interviewee 40)

From the data, we were able to discern a number of ways in which people assess information quality. These included: being sceptical and wary and reading material carefully; comparing it with their existing knowledge; using sites set up by known organisations (e.g. Diabetes UK); disregarding American sites; cross-checking information with books or leaflets; asking their doctor's opinion on the information; looking for the evidence-base; and (paradoxically perhaps) relying on intuition or 'gut feeling' (Nettleton et al., 2005).

Significantly, the 'dangers' of online health were invariably seen as constituting a problem for 'other people' and, without exception, the study participants presented themselves as 'sensible' and 'cautious' users.

It's made the world smaller, but it's not always necessarily good because one person who has had a tin pot idea over the other side of the world, could get a lot of people to think, 'cor yea that's a really great idea' but there's no evidence to back it up. I'm wary I don't just believe everything I read. But then some people do. (Interviewee 10)

In assessing what to trust on the Internet Nettleton et al. (2005) found that people readily draw upon ideologies of professionalism, biomedicine, nationalism and anti-commercialism. Thus indicating that whilst some theorists may be envisaging the demise of ideological and 'discursive knowledge' (Lash, 2002) the practices of lay people reveal that they may in fact endure. Rather paradoxically perhaps, precisely because of the sheer diversity and questionable quantity of information and advice it may be that people are drawing on more conventional and traditional notions of acceptability. Faced with a multiplicity of information and advice sources people may be resorting to traditional ideas about what is and what is not acceptable knowledge. Many would draw on the 'at the end of the day' metaphor to indicate that they would trust their own doctor in particular and biomedicine in general. O'Cathain et al. (2005) have demonstrated how in NHS Direct despite reservations about pressure on the NHS people still trusted it, trusted professionals and trusted this new service because it had an NHS 'stamp'. Thus although people are sceptical at times this scepticism is based within an overall environment wherein trust in professionals and the NHS was still strong. People trust but not slavishly: rather they place trust within an overall context based around their needs recognising that these may be different to those of the NHS and medical professionals on occasion (Hanlon et al., 2005).

Conclusion

The aim of this chapter was to draw upon recent empirical work which has examined lay people's experiences of two innovative health information technologies in order to assess the extent to which the observations and claims made by medical sociologists decades ago have changed. Engagement with contemporary social theory would suggest that transformations are likely to be quite considerable. And indeed there we have observed differences between the findings of today and the second half of the previous century. For example 'patients' come in many more different guises: they may be online health information seekers, telephone health care users and the practitioners with whom they engage are equally as diverse. Both the empirical studies reported in this chapter found that people were informed about health issues and in addition had a strong sense that they had a right to be informed.

However what is perhaps salient about these findings is that many of the features and conceptualisations of health care use identified by Zola (1973),

Stimpson and Webb (1975), Stacey (1988) and their contemporaries still have a validity today. People's pathways to care are rooted in their wider social circumstances, their particular health care needs and, in terms of gender at least, are structurally constrained. In many respects these considerations are precisely what the policy and technological developments are responding to. But somewhat paradoxically there seem to be two processes at work here. On the one hand there is a *growing diversity of health care provision and use,* and yet on the other hand *the norms and values that underpin notions of health care use are concurrently contributing to a reinforcement of caution and conventionality.* Whilst health care users may be engaging with an increasingly complex patchwork of services their conceptualisations of what it means being a good patient guides their help seeking practices. One might even speculate that greater diversity might also result in conventionality. But we as yet have no evidence to support such a fancy.

Nevertheless from our data we see how people make use of ICTs to clarify and assess illness symptoms in order to be a 'good patient' rather than a time-waster and they put these new services within an overall context of the NHS and trust in medical professionals. These avenues are complementary to the broader range of services rather than replacements for them or simply sources of challenge to the medical fraternity. The consequences of this are not dramatic, but we do get a sense that the information gained is contributing to the informal health work that is routinely carried out by people caring for themselves and others. Being better informed can facilitate such work, and provides an important supplement or extension to formal care. But this does not replace the need for medical care, and there was an overwhelming sense in our sample that people do, and indeed want to, trust and rely on health professionals.

Acknowledgements

This chapter draws upon two major studies and so there are a number of people who were involved in the research whom we wish to thank: Lisa O'Malley, Roger Burrows and Ian Watt worked on the e-health project reported here, and David Greatbatch, Donna Luff, Alica O'Cathain and Tim Strangleman on the NHS Direct project: their involvement with the collection and the analysis of the data was invaluable.

Notes

1. This interview and ID number is derived from the data generated from the study by Nettleton et al. (2004).
2. This interview and ID is derived from the data generated from the study by Hanlon et al. (2003).

5
Desperately Seeking Certainty: Bone Densitometry, the Internet and Health Care Contexts

Eileen Green, Frances Griffiths, Flis Henwood and Sally Wyatt

Introduction

Although HRT has long been used by midlife women as a health intervention either for the relief of menopausal symptoms and/or for the prevention of osteoporosis, debate continues about its relative risks and benefits. This chapter examines the ways in which two very different innovative health technologies mediate and inform women's decision making regarding the use of hormone replacement therapy (HRT). Bone densitometry in the Griffiths/Green study and the Internet in the Henwood/Wyatt study both figure in such decision-making practices as women search for certainty regarding both diagnosis and treatment options. How do women make decisions about HRT use in the face of such uncertainty? What information sources do they draw upon? How significant are bone densitometry and the Internet in producing the certainty women seek? What role does the context in which information is found and interpreted play? These questions are addressed in this chapter.

Bone densitometry can be defined as the means by which bone mass measurements are made using dual energy X-ray absorptiometry (DEXA) technology in order to make a judgement about the risk or onset of osteoporosis or 'brittle bones'. Popular myth portrays osteoporosis as an elderly women's disease that starts at the menopause, becoming a crippling degenerative condition that can lead to severe fractures, disability and even death, but that can be 'cured' by HRT (Gannon, 1999). This myth may have arisen from the common confusion between osteoporosis and osteoarthritis (Griffiths and Jones, 1995). The latter can cause ongoing pain and disability, whereas osteoporosis in itself is apparently not painful, rather with osteoporosis a bone may more easily fracture, and this may be painful and disabling. Definitions of osteoporosis or risk of osteoporosis refer to a person's bone density measurement relative to the expected bone density for a person's age and sex (based on bone density measurements of a so-called 'normal' (in the sense of statistically normal, randomised) population). The World Health Organisation's definition of osteoporosis is a bone density reading of at least 2.5 standard deviations below the mean bone density for premenopausal women (Wilkin, Devendra et al., 2001). However, a woman with a bone density between the mean and the level for diagnosis of current osteoporosis may be deemed at risk of osteoporosis in the future as bone density tends to decline with age.

Although bone densitometry continues to be widely used in medical practice, there is debate in the medical literature between clinicians for and against its accuracy in predicting an individual's risk of developing osteoporosis and suffering a fracture (Wilkin, Devendra et al., 2001; Cranney 2003). It is for this reason that it is especially interesting to explore how bone densitometry figures in women's decision making concerning HRT use which, at the time of the studies reported here, was the gold-standard treatment for those considered at risk of osteoporosis.

The Henwood/Wyatt study examined people's health information practices and the significance of the Internet in their wider information landscapes. It is now quite widely argued that knowledge acquired via the Internet, either through health websites, discussion lists or chat rooms supporting online health communities, can empower patients to challenge the authority of medical experts and offer alternative perspectives on health and illness (Hardey 1999; Burrows 2000). However, as with bone densitometry, such claims about the relationship between the technology and associated social practices are contested and, as initial publications from the Henwood/Wyatt study have shown, a number of constraints exist that prevent the emergence of such new patient identities (Henwood, Wyatt et al., 2003; Hart, Henwood and Wyatt (2004)).

We characterise both bone densitometry and the Internet as 'information' technologies, as they both have the potential to provide women with information related to decision making about HRT use. There are, however, three important distinctions to be made between them. First, bone densitometry is always situated in a clinic and relies on health professionals both to produce the readings and to interpret them. The Internet, whilst it can be used in a clinic or doctor's practice, can also be used at home, at work and in a range of other private and public spaces; and the information accessed is not subject to professional interpretation. Second, whilst the timing of bone densitometry use tends to be decided by health professionals more often than by patients and is relatively infrequent, the Internet can be accessed time and time again, as and when users perceive a need for information. Third, unlike bone densitometry, the Internet is a technology that links the user to many diverse sources of health information and it is access to these diverse sources that are seen as opening up possibilities for a challenge to traditional doctor–patient relations.

In the next section, we present and analyse data from both studies regarding the ways in which women attempt to make use of the different types of information provided by these two technologies. We first explore how bone densitometry is used by women and health professionals in both the 'diagnosis' of a problem (osteoporosis), and in the decision whether or not to take HRT. We explore if and how bone densitometry is understood as providing information about the state of people's bones, the need for medical intervention, and the status of this information in decision-making practices regarding HRT. We then examine the Internet's significance as a source of information that can inform health decision making concerning the risks and benefits of taking HRT and, as with bone densitometry, the status of such information in women's wider decision-making practices regarding HRT use. In the final section, we point to the ways in which health professionals and women try to create certainty from the statistical information produced by a medical technology such as bone densitometry, and from the diversity of health-related information, sourced, for example, from the Internet. We also discuss the context in which this information is interpreted, drawing attention to the differences between the clinical context of bone densitometry and the more diffuse context in which the Internet may be consulted.

Reading bones

In the UK, bone densitometry can be accessed by women in a number of ways, including through the National Health Service (NHS), free at point of access, and by women paying directly for the service. However, most women use the NHS, including the women in this study, who were recruited at an NHS clinic. The women had all been referred for the measurement by a doctor, mostly their general practitioners, and so they had already discussed osteoporosis with a health professional. The provision of bone densitometry via the NHS in the UK is uneven, and women's knowledge of the service is generally low so some women may already have spent some time finding out about it and persuading their general practitioner of the need for the test. However, the women talked very little about how they had reached the clinic. It seemed that for them the result of the test overshadowed any other earlier experiences. The bone densitometry result consists of a computer-generated printout including a table and graph of the density measurement (usually hip and spine) in comparison with the 'normal' population. In all cases this printout is interpreted by a doctor. The analysis presented here focuses on women's experiences of receiving the result of their bone densitometry, based on recordings of their consultation about the result or on an interview within a few weeks of receiving the result. In particular, we focus here on women's use of the bone densitometry information in relation to decision making about HRT and so include a little discussion of how this decision making relates to the woman's own life experience. Elsewhere we have shown that for many women making a decision about using HRT takes into account the woman's embodied experience, their perceived health risks, social issues such as how they present themselves in social and work situations as well as advice from experts and test results (Green, Thompson et al., 2002).

In most of the consultations audio-recorded for the research, the clinicians showed the women the results as printed out by the scanner, as illustrated by the following extract:

Doctor: And just to show you this, it's probably easier to sort of eyeball it than sort of discuss it umm, here's the measurement of the lumber spine here shown on this graph and you can see your value is this little box
Patient: Mmmh
Doctor: The blue band is the average result for the population
Patient: Right
Doctor: Okay, so you can see there that your value actually, is just bang on the average
Patient: Good [Sarah]
[This and all names below replace respondents' actual names]

It is interesting that the doctor states that it is easier to use the graphic image as a visual aid to understanding, than it is for him to explain the statistical information to the patient. The clinician's emphasis upon visual images presented in the 'objective' language of graphs produced by the DEXA scan seems to indicate he feels it enables him to communicate. However, we would argue it obscures the subtle and, we would argue, unclear boundaries between what is considered healthy, unhealthy and risky, together with what is a statistically 'normal' bone density reading for the general population (Lauritzen and Sachs, 2001). Furthermore the women fail to understand that it is a population-based measurement rather than a close prediction of their individual risk of developing osteoporosis. Our data from women having bone densitometry indicated that informing women that their bone mineral density readings were above, within or below 'normal', with the aid of a statistically-driven graph which plotted their individual result against the population norm, appeared to encourage them to give excessive weight to this result in decision making. Thus we have a technology-driven calculation assuming an elevated status within the clinical context.

The presentation of this technologically-produced information seemed to leave uncertainty out of the picture. However, in some consultations the doctor took considerable time to explain the meaning of the bone densitometry results and the risks and benefits of medical interventions and lifestyle changes to prevent osteoporosis. They also commented on another form of uncertainty arising from machine error, which occurs where the change in bone density being measured is very small:

> over the last three years they (measurements) fell 5% ... the machine can detect a 5% change, so your bone mineral density is starting to fall ... (Consultation (01) Bone Clinic)

Although there was all this detail in the consultations including uncertainties about the measurement and prediction of outcome, the overall tenor was of certainty (Griffiths, Green et al., 2005). Thus, in this clinical setting, the women heard certainty from the doctors whom they saw as experts necessary for the interpretation of the results. The women took from the consultations a continuing prioritisation of the technological result (as interpreted by the expert) over other considerations. Thus in the consultations there appears to be a co-production of over-interpretation of DEXA scan results and this over-interpretation takes on a pivotal role in the woman's decision making about HRT and other interventions.

In interviews, bone densitometry was discussed as the single most important factor that led women to take medication for prevention of osteoporosis.

> I've just been talking with the consultant now. They're going to put me on to some medicine now which will try and halt the density of my bones getting any worse. [Rachel]

What is striking is that women do not discuss the meaning of the measurement beyond interpreting it as this pivotal point. Although the data from consultations in the clinic included long accounts from the doctors about the meaning of the measurement, most of it was very technical and inaccessible. Eight of the ten women in this particular data subset talked in terms of having a 'diagnosis' of osteoporosis but this seems to be an over-interpretation of the results. Close reading of the interviews would indicate most of the women may have been at increased risk of osteoporosis in the future but did not currently have osteoporosis. We have two cases of women for whom we have both the recording of their consultation with the doctor and an interview: one of these was told in her consultation that her bone densitometry was normal yet during an interview a few weeks later she talks about having osteoporosis. Although the doctor clearly says the bone densitometry was normal, we would argue that it is not surprising that this was missed by the woman as it was embedded in a great deal of detail about future risks of osteoporosis. Here the woman has misinterpreted what she was told. Other women referred to their bones as 'abnormal' and the role of their medication in bringing it back towards 'normal' as measured by repeat bone densitometry. However, in the language of the specialists using bone densitometry, their bone density levels may have been slightly below the average for the wider population but not abnormal in the sense of having osteoporosis.

It is clear from the data that a number of women thought that DEXA scan results provided a direct representation of the health or otherwise of their bones, and therefore a 'window' into the body and, more pertinently, a bone-mass body (Klinge, 1997). One of the interviewees suggests:

> I go [for bone densitometry] once a year and I think you've got to have an awareness of this, I mean these things do happen, and a year is another year on, and I'm not sure quite what's happening inside my body, whether the things are ... You see with bones you don't know whether they getting any weaker or not. [Monica]

This extract suggests a picture or image of a body held up by a skeleton at risk of osteoporosis. The image has been generated by a series of 'bone narratives' drawn from discussions about post-menopause-related osteoporosis risk. Yet the technology is only providing information about one aspect of the complex structure of bone and how it functions as part of the body, that of the fairly simple physical property of its density (the amount of calcium being a key determinant of the density). The following extract is a radiographer talking to a woman while she is on the DEXA scan. It provides a vivid example of the role of the scan as a technological process producing pictures of the bone-mass body.

> *Radiographer:* Yes it's [bone scanner] going all the way through and the picture I get on the screen here is, as each time it goes across your body it produces one line of the area that we're scanning.
> *Monica:* Oh right
> *Radiographer:* And it's just a whole load of jigsaws that build up a picture. It shows me how much mineral is in your bones. I'm doing your hip, we know how much you should have for your age, sex, ethnic origin, what is normal for your age, so just compared with the normal data.
> *Monica:* Oh right, so it's just the mineral content?
> *Radiographer:* Yes, measuring how much calcium [is] in your bones.

It is not part of the radiographer's role in this clinic to interpret the reading: such interpretation falls to a doctor. However, the above extract illustrates the important impact which other health professionals' informal interpretations of the scanning process might have for women, giving an impression of providing a window into the body telling us what is inside and whether this is what should be inside. Understanding bone densitometry as a window directly into the body seemed to contribute to the bone densitometry result becoming pivotal in women's decision making about the need to take HRT or other medication or supplements.

Overall we suggest a co-production by women and health professionals of over-interpretation of the results of bone densitometry, giving excessive attention to the measurement, and so limiting the consideration of other factors such as general health, diet, exercise, how they feel and how they view their health risk which, as discussed above, in other situations women talk about as important (see also Chapter 1 this volume). The apparent certainty conveyed by the use of images, numbers and graphs was reinforced by the sense of certainty conveyed by the doctors and by the understanding women held of bone densitometry as a window into their body.

Bone densitometry features as pivotal in women's ongoing assessment of the state of their bones and the efficacy of the medication being taken. Margaret spoke of her bones having 'improved' as a result of treatment. Rachel spoke of feeling better 'psychologically' once she knew that she had managed to 'stop these things happening', and Monica commented that 'the bone densitometry just tells me where I am each year, whether the drug that I'm currently on is doing any good or not'. In all these cases, it is bone densitometry that has provided them with this reassurance and sense of certainty about both diagnosis and treatment.

We did not find any evidence that bone densitometry has disrupted, challenged or redefined existing social relations in health care (Webster, 2000; Webster, 2002). Rather it appears to have reinforced existing practices of reliance upon 'experts' for the interpretation of medical information. In the Griffiths/Green study, bone densitometry was undertaken in a clinical setting with the doctor interpreting the results. It is possible that the lack of disruption, challenge or redefinition is at least partly due to the positioning of the technology within the hierarchical setting of the clinic, where the use of technology and technologically-produced information serves to reinforce existing power relationships. In the next section we consider information from the Internet, where access is not restricted by the health system.

Going online

In this section, we concentrate on women's perceptions and experiences of the Internet as a means of accessing health information that they think can help to overcome their uncertainties surrounding HRT use. We are particularly interested in the status of the Internet compared to other information sources and in how it figures in doctor–patient relations. We explore the experiences of three women who, for different reasons, all experienced some uncertainty regarding the use of HRT. Each woman wanted to improve her knowledge and understanding about the risks and benefits of HRT and its suitability for use in her own specific circumstances. We present their stories before reflecting on what they tell us about the significance of the Internet in the context of HRT decision making and its wider role as an innovative health technology.

Ruth was very concerned with making the 'right' decision regarding HRT use. She had always thought she would go through the menopause 'naturally', as had her mother and Ruth had thought, 'no, I'm definitely not going on HRT'. However, after having a hysterectomy she felt very differently, that she had no choice, 'because of the situation, I just thought, "well, I'm on HRT" '. At the time of the first interview, she had never looked up anything about HRT on the Internet, saying, 'I deliberately haven't tried to scare myself.' She wanted to be sure about her health and treatment options and, for her, this meant trusting the medical profession:

> I'd have to be so certain that I was right, and I don't think I could be that confident to be honest. I know I have been right about things in the past, in some medical issues, but I would never go so far. I would always trust, rightly or wrongly, the medical care, the medical personnel.

However, her concerns about taking HRT returned, especially after seeing media coverage of the abandonment of a US clinical trial of HRT (Writing Group for the Women's Health Initiative Investigators, 2002) and the development of a lump in her breast. She did, at that point, read widely and search the Internet. However, she found it impossible to find anything that was relevant to her own situation as a woman on an oestrogen-only HRT preparation, who had had both her uterus and ovaries removed. At this point, she returned to her doctor for further insight and advice but did not manage to find certainty there either. She reported her doctor as saying, 'the truth of the matter is that it's a bit of a gamble'. Ruth then hedged her bets and started cutting her HRT patches in half. Ruth told us that her doctor finally agreed she should reduce her dose in line with the amount she had been getting with the halved patches.

Sharon experienced an early menopause (aged 42), ten years prior to the study, and was at that time advised by her doctor to start HRT, warning her that she could 'end up a little old lady' if she did not. Sharon was experiencing hot flushes and aching joints and her children described her as 'the mother from hell'. She knew and was concerned about the breast cancer risk associated with taking HRT but the risk of developing osteoporosis in the future was a bigger worry for her. Sharon was interested in alternative approaches to the relief of menopausal symptoms and had tried some. She had learned about such treatments via the local health food shops and the Internet, in particular, the 'natural health sites'. Sharon felt that an advantage of the Internet was that it could counterbalance the one-sided views of doctors:

> I found that really strange – that most doctors see the advantages of going onto HRT. They seldom point out any disadvantages at all ... on the Internet you seem to find all the disadvantages of being on HRT.

Sharon had a hysterectomy about six months prior to the first interview for the study. The consultant also removed the ovaries at that time because Sharon's mother had had ovarian cancer. By the time of our second interview with Sharon, five months later, there had been a great deal of media coverage regarding the clinical trials of HRT in the US. Sharon told us that she had been aware of this and that she had tried to look for more information online:

> I suppose I had a slight concern when there was all the business in the press about how they now thought that HRT didn't prevent heart attacks and could cause strokes so I was a bit concerned then and that's why I asked [consultant] about it when I went to see him [for new HRT implant]... I did do a bit of surfing but it wasn't very helpful because it said that actually they were still looking at it but nothing had actually – this was in America I believe – and nothing had actually been proved in this country. They were still looking at it so it didn't really get me very far.

On this occasion, Sharon could not find the information or reassurance that she was looking for online. Consulting various sources via the Internet was something Sharon did regularly, and was part of wider and ongoing dialogue between her and the health professionals caring for her. Sharon's ultimate decision to stick with the HRT implant was made in conjunction with her consultant whom she had come to trust. He reassured her that the US trial results probably did not apply to her situation as they referred to the trial of a combined HRT and Sharon was taking an oestrogen-only preparation. At the end of the day, the decision to continue taking HRT appeared to be a shared one, heavily influenced by the fact that Sharon felt 'so well', 'fit' and 'mentally better' whilst on it.

Kathy also had many concerns about taking HRT. She had first been prescribed HRT for a 'hormonal imbalance' when she was 49 (eight years prior to the first interview) and she took it for three years. Like many women, Kathy's ambivalence about taking HRT meant that she had stopped and started again and tried different preparations. She very much wanted to engage with her GP about this but felt that by doing so she had become labelled as neurotic. Kathy read widely and sought to become informed but had no access to the Internet when she was first interviewed in late 2001. By the time of the second interview, eleven months later, Kathy had Internet access, had undertaken some training and had looked up health information on behalf of her sister but not for herself. This seemed strange given her interest in health and her ongoing worries about HRT use. Exploring this further with Kathy, we concluded that her relationship to the health care professionals and her feeling of being labelled as neurotic was actually working against her attempts to increase her understanding about the risks and benefits of HRT use. Kathy's reluctance to search further for information about HRT and ultimately her decision to continue taking it was linked to her perception that, at the end of the day, she was required to comply with her doctor's advice or risk being refused treatment if she developed osteoporosis:

> I was still very anxious about being on HRT and all I wanted to do was to come off it but when you read the literature you're given in the packet it said you must consult your doctor ... and whenever I tried to consult my GP, they didn't want me to come off ... I didn't want to be on HRT, I didn't want to upset him because he told me that the reason he wanted people to go on HRT was for preventative medicine, to prevent against osteoporosis and heart problems, and the way the NHS was going, if people didn't look after their own health, they would become too expensive to treat.

These three accounts illustrate that whilst women attempt to weigh up the advantages and disadvantages of HRT use, risk calculations can prove impossibly difficult. HRT decisions, whether to start, to continue or to stop, are most often made on the basis of women's personal and family histories and their own embodied experiences (Green, Thompson et al., 2002; Green et al., 2004). The decisions are reviewed in the light of changing embodied experiences and any new information sought and found as a result. The Internet does figure in women's decision making regarding HRT use but, unlike bone densitometry, it has no enhanced status in these decisions. Rather, it appears as one of many information sources used by women, supplementing rather than substituting for more traditional sources. Furthermore, these stories show that the Internet does not affect health information and decision-making practices in any universal way, neither to reinforce the traditional authority of health professionals nor to empower patients. Different scenarios are possible.

Internet use has increased at the same time as the shift towards a more pro-active role for patients has become part of government policy and popular discourse but it would be wrong to conclude that the technology caused this social change. The stories presented here illustrate well how the Internet enters into existing social relations, both shaping and being shaped by them. The Internet figures in all three women's accounts of their HRT decision-making but in very different ways. For Ruth, the Internet figured as something that would simply 'scare' her and she was not a keen user; for Sharon, it figured as a source of alternative information about treatments for menopausal symptoms in the early days of her HRT decision making; and for Kathy, it came to symbolise a threat to traditional doctor–patient relations. Both Ruth and Sharon managed to incorporate limited Internet use into their information and decision-making practices in which their doctors also played a key role. For Kathy, the story was very different. Existing relations with her GP were poor and she felt uncomfortable about asking too many questions prompted by her own information searching. In this case, the Internet was perceived as exacerbating an already difficult relationship and Kathy completely rejected its use for her own health management.

Medical information technology and information technology for health

This chapter has examined the role of two innovative health technologies, namely bone densitometry, a medical technology that provides information, and the Internet, an information technology that can be used in relation to health. Webster has defined innovative health technologies as technologies that occupy a 'contested terrain' and disrupt, challenge or redefine existing social relations and practices in health care (Webster 2000; 2002). Thus, the emphasis is on the relationships and practices rather than the novelty of the artefacts themselves. This has proved to be relevant here as both bone densitometry and the Internet have relatively long histories although individual women may experience them as new. The data presented above demonstrate how technologies are being used as a source of information for health-related decision making. This is an example of a much broader debate which Webster (2002) refers to as 'medicine as information', arguing that medicine is being increasingly informaticized alongside mounting levels of uncertainty about the clinical relevance of the information (Fox, 2002). The interesting questions in relation to bone densitometry and the Internet then become what relations and practices are being referred to, what kinds of contestations exist and what disruptions, challenges or redefinitions of those relations and practices are taking place? A key area of contestation examined in this chapter was about how far traditional knowledge/power relations between patients and health practitioners are challenged or redefined through the use of technologically-produced information, whether generated by medical technologies such as bone densitometry, and interpreted in a clinical context, or found by women themselves via the Internet in a variety of settings. This chapter examined these questions in the context of women making decisions about HRT.

Bone densitometry is understood by many women as producing knowledge about the state of their bones that helps to reduce uncertainty regarding the use of HRT. These women had set out on a process of negotiating the health service to gain this information. The use of the technology to provide what most women assumed to be an accurate 'window' into their body seemed to influence strongly the dialogue in the clinical context. The combination of a technologically-produced measurement communicated to women via a medical expert in a specialist clinic resulted in women feeling confident that they knew about the state of their bones and the need for, or efficacy of, treatment. Both patient and health professional engaged in attempts to enrol the technology in the practice of reducing the uncertainty associated with women's future health status, despite the health practitioners' knowledge and comments which contradict this. Bone densitometry appears to be powerful in the technologically-mediated process of risk prediction and the search for certainty. Unlike information from the Internet that was interpreted within the context of the woman's embodied experience, the information from bone densitometry seemed to be given priority over other considerations. Moreover, traditional patient-provider relationships were reinforced as practitioners continue to perform the role of experts required for the interpretation of information produced by medical technology. It is unclear whether the extent of this lack of challenge to the traditional patient–doctor relationship is due to the technology being situated within the NHS and within a clinical context.

The Internet was understood by some women as a source of information about bodily changes associated with midlife and about the various treatment options, including HRT. Two women in the study appeared to access health information in a relatively unproblematic and direct way, and thought this would empower them in their decision making. Interestingly, despite being able to access the Internet, their inability to find information exactly relevant to their circumstances was a factor that led them to question their health professionals. However, another woman felt unable to challenge her health professional. The Internet as an information technology relevant for health did not take on a pivotal role in women's decision making about HRT as bone densitometry did. This may in part be because the Internet is accessed outside of the clinical context and becomes only one element in a more widely-based consultation with a health professional about HRT. One question raised by the analysis in this chapter is about what would happen if the Internet was consulted in the context of a clinical setting. If that happened, would women be directed towards medically-approved sites and would we then see, as with bone densitometry, a co-production of certainty?

Conclusion

In seeking to use both bone densitometry and the Internet the women were seeking information on which to base decisions about taking HRT. Uncertainty can be very difficult to live with, and many of the women seemed to want to use the technologies, the information they generated and the apparent expertise of health professionals in a search for certainty. Although both bone densitometry and the Internet are established technologies, both are often experienced as 'new' and 'leading edge'. As a result, perhaps they still carry with them the seductive promises of technological progress (Brown, et al., 2000) and it is this which leads many women and their doctors into looking for certainty where there may be none to be found. The two technologies are uncertain in relation to health but in very different ways. The uncertainty for bone densitometry derives at least in part from the limitations of statistical probability: an accurate estimation of risk for a population provides no certainty for an individual. The information from the Internet shares this limitation. However, the uncertainty of the Internet also derives from the multiple and contested sources of information available there.

With bone densitometry, the technological mediation and its use in the clinical setting leads the information it provides to be overvalued and become pivotal in women's decision making about HRT. With the Internet, the women started out with the expectation that it would provide them with value and decisive information about their individual health situations. However, their experience was disappointing as it failed to provide them with such specific information and they returned to talk to their doctors.

Many women's decisions remain heavily mediated by 'expert' interpretation, with little disruption of established knowledge/power relations despite the apparent potential for information to empower health care users. There was limited evidence of any fundamental transformations in patient–professional relationships resulting from the greater availability of health information, however produced. The potential remains, however, for women to develop sufficient experience in seeking and interpreting health-related information from the Internet and when provided with information from medical technology such as bone densitometry, of interpreting this for themselves, to enable them to deal with the uncertain and provisional nature of different sorts of health information.

6
Telemedicine, Telecare, and the Future Patient: Innovation, Risk and Governance

Tracy Finch, Carl May, Maggie Mort and Frances Mair

Introduction

This chapter charts the problematic interactions around a particular field of telecommunications in medicine, or, perhaps, a field of medicine in telecommunications. In it, we explore how 'telemedicine' has failed to normalise as part of everyday clinical practice in British health care, how attention has shifted first to a mode of practice which we call 'telehealthcare' and how this has been displaced by a new mode of practice – *telecare*. Before we begin our account, it helps to offer some definitions, for throughout this paper we use three terms to characterise the developments with which we are concerned.

(i) *Telemedicine* refers to synchronous or asynchronous electronically mediated doctor–patient interaction aimed at diagnostic or management planning for patients.
(ii) *Telehealthcare* refers to a wider and more diffuse set of systems, often employing nurses, that develops and delivers advice and treatment management, where participants are geographically separated.
(iii) *Telecare* refers to remote surveillance (chronic disease management 'solutions') of illness trajectories and to collect information about symptoms or physiological parameters (for example 'peak flow' in asthma, or blood sugar levels in diabetes).

The longstanding assumption of their clinical and policy proponents has been that Information and Communication Technologies (ICTs) will 'revolutionise' health care delivery by transforming relationships between patients and health professionals (Haux, Ammenwerth, Herzog and Knaup, 2002; Ball and Lillis, 2001). In relation to this it has been argued that new health technologies are redefining medicine, health and the body (Webster, 2002), and that ICTs are part of this reconfiguring (Wajcman, 2002). Telemedicine and telecare enable remote medical and health care delivery by linking patients and health professionals electronically, and seem to be an important part of this shift because of their promise as new forms of health care provision (Lenaghan, 1998) They have attracted considerable interest amongst policy makers (Department of Health, 2002)and health professionals (Wootton, 1997) as a solution to problems of access to health care services and as tools for the management of demand, especially for specialist services.

These developments are supported in policy initiatives, that advocate both the increasing use of ICTs for improvements in the efficiency of health services provision, and which place emphasis on increasing responsibilities for patients to engage in self-care by mobilising electronic resources (Kendall, 2001). It is argued that these technologies permit not only the monitoring of individual patients' conditions at a distance for the purpose of management and treatment, but have important broader implications for the practice of medicine and health care, enabling new forms of practice that are based on the collation, management and exchange of large quantities of patient data (May, Finch, Mair and Mort, 2005). Thus, beyond the immediate implications for clinical practice and interactions between patients and health care professionals, developments such as telehealthcare and e-health raise important questions about changing relationships between the patient, the health care profession, and the State. In particular, many argue that increasing use of such technologies, by providing greater access to 'information', will create unprecedented opportunities for the *empowerment* of patients and citizens (Gann, 1998; Lenaghan, 1998).

An increasing emphasis on information as the basis for medicine is evident. Analysing policy documents advocating increasing use of ICTs to transform health care in the UK, Klecun-Dabrowska and Cornford (2000) argue that policy statements represent information as something that can be easily captured, understood and transformed to achieve the goals of the NHS (e.g. improved efficiency), and that these 'fantasies' of information are also those of telehealth, serving to remove meanings of telehealth from a local, organisational and human context. Their analysis highlights the complexity of issues involved in telehealth, and they warn that 'telehealth is not a given, even within the policy debate, it is a concept that has to acquire, develop and sustain meanings'. Empirical research demonstrates the complexity of achieving this in practice (Sicotte and Lehoux, 2003).

Telemedicine has been approached largely as an *innovation* in the application of generic ICTs to the provision of health care (Rappert and Brown, 2000; Whitten and Collins, 1997). In practice, such initiatives tend not to become 'normalised' (May, Harrison, Finch, MacFarlane, Mair and Wallace, 2003), for they often fail to progress beyond trial and demonstration projects to become part of routine health care delivery (Clough and Jardine, 2003; Bashshur, 1997). Clinical researchers in this field have approached the problem largely in terms of barriers to 'implementation' (Wyatt, 1996), thus eliding the complex social and political contexts in which relations between new technologies and their users are played out (Mackenzie and Wajcman, 1999). There is a developing body of research which seeks to understand telehealthcare as complex socio-technical systems (Aanestad, Rotnes, Edwin and Buanes, 2002). Studies that have focused more contextually on the *processes* of telehealthcare services reveal major problems of integration of telehealthcare systems into local service delivery contexts and networks (Barlow, Bayer and Curry, 2004; Bayer, Barlow and Curry, 2004) and existing professional roles and relationships (May, Gask, Atkinson, Ellis, Mair and Esmail, 2001; Mair, Whitten, May and Doolittle, 2000; Mort, May and Williams, 2003; Lehoux, Sicotte, Denis, Berg and Lacroix, 2002), often requiring a complex reworking of professional–patient interactions that take place in clinical encounters mediated through such systems (May, Ellis, Atkinson, Gask, Mair and Smith, 1999; Miller, 2002). Problems concerning the implementation and integration of telehealthcare services have been further exacerbated by struggles over the production and evaluation of evidence about clinical and cost effectiveness (May and Ellis, 2001).

For those health professionals struggling to adopt and adapt to new ways of delivering health care, challenges to established routines and practices generate feelings of uncertainty and raise questions about potential risks to patients and concerns about personal liability as practising professionals. The appraisal and management of risk and uncertainty by users of telemedicine and telehealthcare systems has however, received relatively little attention. In the clinical context, the problem of risk management has been approached either through the development of clinical safety guidelines and protocols or through legal consultation (Stanberry, 2001). From the research community, attempts to prove scientifically that such systems are safe, and clinically effective, have predominated (Hailey, Roine and Ohinmaa, 2002); however, scientific reviews examining the quality and outcomes of such research studies conclude that there is not yet a strong evidence base for telehealthcare (Urquhart, Currell, Lewis and Wainwright, 1999; Mair and Whitten, 2000; Whitten, Mair, Haycox, May, Williams and Helmich, 2002). Evaluation studies also fail to adequately represent the perspectives of both patient and professional users of telehealthcare systems (May, Mort, Mair, Ellis and Gask, 2000; Roberts and Rigby, 1998), thus the ways in which users perceive and negotiate the potential risks attending these technologies remains poorly understood. There is some evidence, however, that both providers (Mort et al., 2003) and recipients of health care services facilitated by ICTs engage reflexively with new systems of practice in order to accommodate potential risk .

The research literature around telemedicine and telehealthcare has thus far failed to adequately address important questions about the increasingly complex networks of policy and practice on which these developments are predicated. This chapter therefore examines the links between these services and their points of public and private accountability, and the reconfiguration of patient-hood that they seem to promise. Based on a recent study of policy and practice around telemedicine we critically examine ideas about risk, innovation and governance as these are apprehended by patient advocates, clinicians, NHS managers and policy actors. Since 1997, we have undertaken a series of linked studies that investigated the development, evaluation and implementation of telemedicine and telecare systems. In this chapter, we draw on a substantial body of empirical data from in-depth qualitative interviews with key informants representing the views and perspectives of a wide variety of actors, as well as data collected from attendance at key seminars and conferences within the field. Although our research covers the broad context of telehealthcare in the UK, we have also undertaken case studies of teledermatolgy, telepsychiatry, home telecare and internal medicine using telemedicine systems. In this chapter, our analysis is focused on the ways that telemedicine is configured in relation to ideas about innovation, risk and governance, and in changing relationships between patients, the NHS and the State.

The 'disappearance' of telemedicine

The most striking feature of our work is that 'telemedicine' as it has been conventionally understood by its clinical and policy champions is *disappearing*. As indicated by one of our respondents:

> 'Well telemedicine is a word I think is on it's way out, in fact it probably already has been replaced by the word e-health which describes something broader than telemedicine [...] maybe one day we'll just say e-health.' [Int-10]

Electronically mediated doctor–patient interactions (intended to deliver diagnosis and medical management), are being rapidly displaced by applications that involve a wider range of staff (mainly nurses) utilising systems explicitly intended to manage the routine trajectories of chronic diseases. Diagnostic services in dermatology have tended to provide advice between clinicians and to cope with a low volume of patients. The exception to this was a private sector supplier, which has recently gone into liquidation.

The 'disappearance' of telemedicine stems from several factors. In the last decade, the struggle for its clinical and policy champions has been to try to channel longstanding policy commitments to modernising informatics into resources for specific telemedicine services. This has proved steadily more difficult as contests over the development of the national Electronic Health Record and its associated systems (notably the national 'Connecting for Health' programme) have absorbed the attention of policy-makers and IT managers in the NHS, and also absorbed much potential funding. Local clinicians and managers have been resistant to telemedicine until the practical implications of this national programme (and especially questions of systems integration) have been resolved. At the same time, the growth of call centres (NHS Direct, NHS 24) and web-based systems that mediate between the material resources of the NHS and its patients have focused attention on services perceived to be highly effective. Telemedicine's champions have therefore sought to co-opt and harness a discourse of more effective *management control* over health care delivery.

Telemedicine ceases to be 'innovative'

From the beginning of the 1990s, 'telemedicine' was promoted and enacted in the National Health Service as an 'innovation'. Yet it is no longer seen in this way by many of its users. Telemedicine was framed in early policy statements (National Health Service Executive, 1998) as 'new', 'exciting', and as having the potential to radically transform medical practice. The Department of Health urged clinicians to embrace telemedicine as a new way of working. Much activity within this emerging field focused on systems' research and development (R&D), with an emphasis on 'proving effectiveness'. Framing telemedicine as an 'innovation' fitted with emergent modernisation agendas, but it also made explicit the need for rigorous (quantitative) evidence to demonstrate its clinical safety and effectiveness. This evidence has been hard to come by (May, Mort, Mair and Williams, 2001). Focusing on evidence production meant that integrating telemedicine into clinical services was difficult, for the organisational complexity of this task extended well beyond the development and evaluation of hardware. It appears that many tried and failed to achieve a telemedicine service that was workable, often because they underestimated this organisational complexity (Finch, May, Mair, Mort and Gask, 2003).

If telemedicine as an innovation is 'disappearing', then telehealthcare and telecare – and 'ehealth' – have almost completely displaced it. This represents a clear shift in how telemedicine is 'framed' as innovative or otherwise, as illustrated by one of our clinical respondents:

> 'I think we've suffered from jargon and I think we've suffered from the jargon of innovation. Innovation is a great word, you know it's used for all sorts of things and became very popular in the sort of nineties er maybe the late eighties [...] and er was a buzz word used for everything. I don't think teledermatology is innovative particularly, I think it's a slightly different route for doing the same thing.' [Int-19]

This configures a shift from *clinically focused innovation* to *organisational change* in service delivery. The technologies and processes on which telemedicine was predicated are now conceptualised as a 'tool' for facilitating organisational change in service delivery. Equipment designers have followed this move, shifting from developmental 'hybrid' systems designed to mediate between the doctor and patient in specific clinical situations, to generic ICTs that can be widely distributed (telecare systems), while service developers have shifted attention to larger scale organisational changes in service delivery utilising call centres (e.g. NHS Direct). Clinical champions of telemedicine now focus on it leading to change in service delivery, downplaying the previously 'innovative' qualities of systems themselves. Consistent with this view, 'champions' of technology developments within the NHS are now often identified as senior managers seeking to control costs of chronic illness, rather than clinicians seeking to innovate around diagnosis and management.

Applied to dermatology, telemedicine is not so much about controlling the costs of chronic illness as providing low intensity clinical advice. Elsewhere (May et al., 2003), we have suggested the notion of *normalisation*, the routine embedding of *locally* framed systems in service provision, as a way of understanding the struggles that this involves. Teledermatology highlights several important features of innovation in telemedicine and telehealthcare. In our work, we have seen a steady diminution of expectations of 'innovation' in the field of dermatology, one that is – on the face of it – ripe for teleintervention. Apart from one commercial provider and a government sponsored co-ordinated service, a further six localised teledermatology services have achieved some degree of normalisation. These services have evolved from earlier teledermatology models (predominantly consultant-led diagnostic services) to include clinics run by nurses or GPs using store-forward systems for digital image and clinical history transmission; or GP to consultant store-forward systems for triage/diagnosis/opinion. None of these deal with large numbers of patients (the most productive had seen around 400 since its inception in 1998).

These 'normalised' services appear to have coalesced around particular local clinical networks and service delivery needs, rather than specific systems. In one of these services, a commercial provider had been contracted to meet locally defined service requirements rather than a 'one size fits all' service. Key features of normalised services are: (i) that they emerged out of service reconfiguration rather than experimental research; (ii) monitoring of the service is characterised as audit rather than evaluation; (iii) they permit the front end provider (nurse or GP) a degree of flexibility and autonomy in the use of the hardware; (iv) they are viewed as adjuncts to existing 'traditional' services; and (v) they have all become part of routine service delivery with teledermatology situated as one of several options available to doctors and nurses. Clinicians working within these services talk of the reorganisation of services, rather than the technology, as innovative. Crucially, these services are – perhaps deliberately – sited outside of the domain of R&D.

Telemedicine and telehealthcare: shifting notions of risk

The perception and management of risk has been an important issue in the development of telemedicine. Archived data demonstrates significant concerns about the clinical safety and accuracy of telemedicine as a substitute for face-to-face medical consultation, and the potential medico-legal ramifications of using telemedicine. The concerted research and evaluation effort that accompanied early telemedicine development was itself a response to concern about risk. At the same time, concerns about the 'risk' of using telemedicine impacted on whether specific developments were seen to 'work' and the extent to which they could be evaluated. 'Risk' at this time was largely constructed as a 'clinical' problem – often related to the potential for misdiagnosis – and strategies for its management were based in clinical service protocols.

With the emergence of telehealthcare, telecare and, latterly, ehealth, the concept of risk in relation to technologically mediated health care provision has become much more diffuse. Participants in our studies discussed an array of 'risks', including clinical (misdiagnosis); social/interpersonal (interactional); personal/professional (liability/role change); technical (failure); and organisational (poor integration). This highlights the multifaceted nature of risk in this context.

For telemedicine, clinical risk remains important but is viewed by those engaged in it as manageable in practice. An important theme in our recent data is that ideas about risk are being reframed in relation to the question 'what is different about telemedicine?' One of our patient advocacy respondents explained that:

> *'there's risk in everything, there's risk in whether a doctor's doing an operation and looks at somebody's appendix and decides not to take it out when it should be taken out, I mean we have to appreciate health and medicine, and the treatment of health, it's a very very risky business. No extra risk as long as it's good medicine with good communication between professionals.'* [Int-03]

This reflected a view that medicine always involves risk and uncertainty and that health professionals need to judge whether any system of practice – technologically mediated or otherwise – permits a sufficient clinical assessment of a patient. It also reflected concerns about potential organisational 'risks', such as security of data and confidentiality of records, issues that were often raised by respondents as potential risks yet at the same time dismissed by them on the grounds that paper-based records and data storage procedures are equally (or perhaps more) prone to security lapses and loss of records.

There are also those working within the field of telehealthcare – several in our recent interview samples – who characterise concerns about risk expressed by those in the field, as a means of 'resistance' to the change in work practices required to operate successfully within these systems. Such views reflect the distinction between perception and management of risk in this context, since those who are enthusiastic about telehealthcare seek out ways of addressing the potential risk of using telehealthcare rather than dismissing it on the basis of it being 'too risky'. Those who focus on management strategies do not tend to see risk as a problem for telehealthcare.

Our data indicates two key ways in which perceived risks are resolved – through evaluation (or 'audit') of services, and by clinical judgement. Evaluation and audit are perceived as important conditions of the introduction of telehealthcare, though many feel that over-emphasis on formal approaches to research and evaluation – and especially randomised controlled trials – by the medical community have impeded progress in telehealthcare developments. Those using telehealthcare state that they are less concerned about *clinical* risk because clinical judgement emphasises caution. Likewise, professionals concerned about threats to interpersonal interaction posed by the telehealthcare system draw on their professional expertise to accommodate the perceived limitations of the technical system. It is important to emphasise that where telehealthcare systems are seen by their users to 'work', they are characterised as *alternatives* rather than substitutes for conventional services – thus concerns about risk can be accommodated if the health care professional can choose between the use of telehealthcare and the traditional service. The value of this as a principle for telehealthcare development was identified independently by members of a 'Citizen's Panel' that we recruited to discuss issues around telemedicine and suggest principles for its development (Mort and Finch, 2004).

Emphasis on clinical risk and liability is particularly evident in teledermatology. Although early studies supported the reliability of diagnosis and management between teledermatology compared with face-to-face interaction (Eedy and Wootton, 2001), concerns about misdiagnosis and liability remain. However, as described previously, clinicians involved in practising teledermatology as part of routine service delivery manage these concerns as a matter of *clinical judgement*. Simply put, if a diagnosis or management recommendation cannot be confidently made on the basis of the data provided through the teledermatology system, the patient is given an out-patient appointment with the consultant. Respondents representing associated patient advocacy groups, generally expressed confidence that potential risk will be appropriately managed in this way. For example,

> 'I don't think there are any particular risks, not least 'cos I have great confidence in the the carefulness of doctors and at all [...] I believe that if there was any doubt they would say I'm sorry I really can't make a judgement on the basis of this photograph, I need to see the patient.' (PL, Dermatology Patient Advocate)

However, the Citizens Panel emphasised the need for safeguards to be built into systems of technologically mediated care, such as adequate training in the use of new systems, and ongoing assessment and evaluation.

Rethinking the *purpose* of teledermatology has impacted on how risk is conceptualised. 'Normalised' teledermatology services have moved away from the initial emphasis on diagnosis, and are viewed by their proponents as a tool for triage, diagnostic advice, management and patient education, with a relatively small diagnostic component. As this consultant dermatologist explains:

> 'I think it's jumping to diagnostics conclusions em is a risk so that's why, we know, that diagnosis is very much secondary it's about managing a known diagnosis so that both, you know the people at both ends of the system know the diagnosis eh all they're doing is, is talking about the management.' [Int-28]

In some cases, these services are explicitly 'informal', existing as a fast route for general practitioners to access a consultant opinion but without replacing the formal referral system or shifting responsibility for the patient away from the general practitioner. In one service, teledermatology is seen purely as a back-up tool for nurse-led dermatology clinics. A service that operates in a remote location uses teledermatology to quickly develop a management plan, even in cases where the consultant will need later to see the patient in person, so that the patient can receive some treatment within the interim period. By taking a wider view of the purpose of teledermatology and its role within a broader system of dermatology services, these teledermatology services are shifting emphasis away from *diagnostic accuracy* where concerns about the risks involved in practising teledermatology have previously focused, thus making risk *manageable* within this context.

Telemedicine, governance and accountability

Applied to health care, ICTs are increasingly enabling shifts in the form and scale of governance and accountability in service provision. As developments have shifted from *telemedicine* towards *telehealthcare* and *ehealth*, different forms of governance have been evident. Generally, we find that, in the UK, there exists no specific formal structure for governance in this context, and that patients and citizens particularly remain absent from decisions about service configuration.

During the development of *telemedicine*, the major governing influence was *evidence* – the demand for robust data about the safety and effectiveness of what were perceived to be new medical practices. Although such research was initially believed to facilitate the development of telemedicine, many respondents said that it actually inhibited integration of telemedicine into clinical practice because of tensions between the requirements of building an 'evidence-base' and the practical requirements of clinical practice within the NHS. They have asserted that government policy has impeded the development of telemedicine in the UK, by initially advocating its widespread adoption, then subsequently drawing in specific funding and focusing instead on generalised *modernisation*. Current policy then, is seen by our respondents to favour more generic widespread applications of ICT rather than the specialist clinical applications of telemedicine that were previously considered to be 'innovative'.

As the applications of ICTs to health care provision in the NHS have moved away from telemedicine, new forms of governance have emerged. The use of ICTs is now being directed more towards systems of 'decision support' (for example, NHS Direct and many telecare applications) that draw on evidence-based medical knowledge to facilitate *consistency* of practice. As this respondent explains:

> *'then there's benefits in terms of getting the right expertise in the right place […] I think that people haven't paid enough attention to the whole movement of er, the possibility of making the whole of health care much more protocol driven you know, with decision support and so on and therefore […] setting up […] standard treatments.'* [Int-09]

In such systems, *collective knowledge* is emphasised over *individual experience*. Views about increased governance of practice in this way are likely to be mixed, however several interview respondents report that it is seen in a positive light. For example, NHS Direct requires staff to demonstrate *accountability* by explicitly justifying professional judgements and decisions, and this can also be understood as protection in the event of allegations of negligence or malpractice.

Our work has also suggested that although telehealthcare has important public implications, wider accountability to citizens remains absent. Telehealthcare and telecare are increasingly advocated as new ('modernised') forms of service *delivery*, but emphasis is on change in the *location* of services and the *mode* of accessing services, whilst the services themselves are presented as essentially the same. Although patients receiving telehealthcare services are often invited to comment about their experience (formally or otherwise), our data reveals an almost complete lack of consultation between the public and providers about the configuration of services. In many instances of telemedicine, telehealthcare and telecare, interview respondents from various backgrounds argue that such services are alternatives rather than replacement services and that patient choice is (or should be) retained.

Questions around accountability and telehealthcare are important, because although such services are reported to *increase* access to health care, they are also a means of *controlling* access – for example, as a triaging device at the point of referral or initial contact. The majority of respondents also express the view that telehealthcare and ehealth, by shifting health care provision away from secondary care, is enabling more patients to 'self-manage' and take on greater responsibility for their health care. Although this shift is viewed positively (even as necessary) by many, others are sceptical about its desirability and workability.

Innovating 'medicine': innovation, risk and governance

Our account of the shift from ('revolutionary') 'telemedicine' to (routinised) 'telehealthcare' and 'telecare' demonstrates the interplay between notions of innovation, risk, and governance (see Table 6.1). Here, the way in which telemedicine was constructed as an *innovation* had direct implications for what was required for its development, implementation, and sustainability, frequently impeding its routine embedding in service provision. Constructing a technology as *innovative* inherently emphasises *risk*. As a system enters practice, the 'gloss' of innovation appears to dull, and the balance between effort and reward shifts. *Risk* must be seen as manageable for a service reconfiguration to become accepted as normal practice, and this is demonstrated by the teledermatology services described in this study. As developments and services have shifted away from telemedicine towards telehealthcare and ehealth, new forms of *governance* have emerged in which increased monitoring and control of the practices of both health professionals, (through computer-driven protocols and built in audit systems) and patients (through increasingly localised self-management of chronic illness) is evident. However, despite policy statements supporting increasing involvement of the public in decision-making about health care services, patients and other citizens seem excluded from structures and processes of accountability.

Table 6.1: Telemedicine and telehealthcare: Innovation, risk and governance

	Early applications: *Telemedicine*	*Emerging applications:* *Telehealthcare and telecare*
Innovation	Emphasised and celebrated Focused on the technical system	De-emphasised – threatens implementation Focused on service delivery
Risk	Central and problematic Clinical and technical	Manageable, facilitated by protocol Diffused/diluted
Governance	Policy advocates telemedicine Scientific and evaluative *evidence* as major form of governance	Policy advocates generic and widespread use of telecommunications technology ICT-based systems designed to promote consistency of practice and demonstrate accountability
Patients	Experimental subjects Prioritising access Absent from decision-making about services	Routine patients Prioritising access Informed and self-managing Absent from decision-making about services

Innovating health care: reconfiguring patients?

Ideas about the political identity of patients themselves change in relation to telehealthcare systems. The notion of the changing patient – from a role traditionally characterised as passive to one ascribed as 'informed', 'expert', 'self-managing', 'activated' and as 'having responsibilities' – pervades responses from interviewees and public speakers that have contributed to our analysis. This shapes local policy and managerial decision-making about how new technologies can be used to modernise health care. Telemedicine and telehealthcare are justified by the presumed preferences of patients for faster access to local services, and the potential for greater 'choice' about modes of access. Telehealthcare is seen to offer ways to achieve these priorities, and on this basis is presumed to be welcomed by patients and citizens. In a keynote address, one of telehealthcare's policy champions (Preston, 2001) drew these preferences and solutions together to present an account of 'modern' telemedicine as 'patient-centric'. However, telemedicine and telehealthcare have implications for patients, and for their relationships with health professionals and the NHS, that go beyond issues of access, and the trade-offs that patients are willing to make against various aspects of health care services are assumed rather than known. Data provided by the Citizen's Panel (Mort and Finch, 2004) conducted for this study illustrates the complexity of the preferences and values that citizens hold for the ways in which services are developed and delivered.

Conclusion

In this chapter we have demonstrated the various ways in which technologies and their users get configured in policy and practice through analysis of the development of telemedicine in the UK. Constructing ICTs as 'innovative' presupposes a set of conditions that can actually make the systems of practice upon which they are based unworkable in a service delivery context. Firstly, this configuring as 'given' rather than as part of ongoing design leads to the neglect of the broader socio-technical network in which technologies are placed. Secondly, it serves to emphasise concerns about risk, safety and responsibility, which require a certain amount of flexibility in practice (and trust in professional judgement) to become manageable.

Finally, constructing telemedicine as innovative requires that new and appropriate forms of governance be negotiated and understood by those who are expected to engage with new systems of practice. This has posed problems for professional users of telemedicine systems, practising in a context of increased uncertainty and professional responsibility. Our study also highlights the extent to which citizens are *not* empowered to engage policy and practice developments in telehealthcare, and thus the ways in which information technologies can facilitate the empowerment of patients remains to be seen.

The ways in which telemedicine and telehealthcare technologies have been constructed as an 'innovation' thus have important implications for whether or not they have the potential to become a 'normalised' part of health care delivery. As the success of a small number of teledermatology services demonstrates, it seems that (to concur with Suchman, 2002), to be truly 'innovative', some technologies must literally 'disappear' amongst the creative negotiations or local improvisations of everyday practice.

Acknowledgements

This chapter is based on the research project 'Telemedicine and the "Future Patient"? Risk, Governance and Innovation', funded by the Economic and Social Research Council as part of the Innovative Health Technologies Programme (ref L21825 2067).

7
Patient 'Expertise' and Innovative Health Technologies

Katie Ward, Mark Davis and Paul Flowers

Introduction

This chapter addresses expertise and health technology via two case studies: the online pro-ana movement and transitions in the treatment of HIV. In different ways, each case study sheds light on the construction of patient/consumer expertise. Before presenting the two case studies, we first outline the notion of the 'expert patient' as it is depicted in recent policy frameworks and consider several critiques of the idea of patient expertise. We then consider how patient expertise is taken into forms of innovative health technology, in particular, Internet-based health care and rapid developments in the diagnosis and treatment of chronic illness. The case studies of the pro-ana movement and HIV that follow help deepen the analysis and we conclude the chapter with a summary that draws out some of the general perspectives on expertise identified in the cases.

The 'expert patient'

Strategies to modernise the NHS include the 'expert' patient initiative (Department of Health, 2001). This programme links 'patient expertise' to ideas of 'empowerment', a 'better quality of life', 'self esteem' and a 'user led NHS' (Department of Health 2001; Neilson, 2003). Expert patients, according to this view, manage their own illnesses and conditions by developing knowledge relevant to maintaining health and countering illness. It is highlighted in the recent Wanless review that processes involved in allowing self care are dependent on patients' or consumers' 'health literacy', which includes fostering the ability to understand instructions and information leaflets, evaluate relevant health related information and effectively interact with health care systems (Neilson, 2003).

Similarly, in a report for the Association of the British Pharmaceutical Industry (ABPI) Illman (2000) establishes a framework for the emergence of the 'informed patient'. According to the ABPI, the informed patient initiative will transform the doctor–patient relationship from a 'professional led' interaction to a 'doctor–patient partnership', in which informed patients ensure that treatments are appropriate to their individual needs.

The new media and the expert patient/consumer

Facilitating the emergence of patient expertise has been the rapid growth in web based health related information. Numerous web and email based forums offer information and discussion of health care; and consumers use websites to research their own conditions and health care and make decisions surrounding their treatment (Bessell et al., 2002). Concerns remain over the reliability and validity of web based information, but as part of the development of an 'expert patient culture', organisations such as the ABPI and the World Health Organisation (WHO) have developed guidelines on how to interpret, assess and use web based information.

Lupton (2002) and Henwood et al. (2003: 597) suggest that the growth in health information availability has played a role in allowing the transformation of the 'patient' into a 'reflexive consumer', where active decisions are taken concerning treatment procedures. This has implications for the understanding of patient-consumer and lay expertise. More specifically, the notion of 'patient-hood' and the idea of the expert patient become questionable as the development of lay expertise and knowledge about a condition allows a shift from patient-hood towards reflexive consumer (Fox et al., 2005a, 2005b; RPSGB, 2004).

The emergence of Internet-based pharmacies and other web based outlets selling pharmaceutical drugs, including 'lifestyle' drugs such as Sildenafil (Viagra) and weight loss drugs, such as Orlistat (Xenical) also contributes to the active consumption of health related products (Fox et al., 2006). The increasing availability of health related information and pharmaceuticals contribute to the emergence of an environment where health matters both permeate and define culture. This burgeoning 'health culture' or the medicalisation of culture is illustrated in the recent redefining of some treatments, such as Sildenafil and Orlistat, as lifestyle medicines or drugs (Gilbert et al., 2000). Lifestyle drugs are used for 'non-health' purposes (Flower, 2004), or as Ashworth et al. (2002) state 'a pharmaceutical product characterised as improving quality of life rather then alleviating disease' (236).

Treatment innovation and the expert patient

At the same time as applications of the Internet have helped push forward the idea of patient expertise, innovations in the management of chronic disease have also made much of the role of the expert patient. This kind of approach to chronic illness builds on a traditional public health idea of primary and secondary prevention and notions of self-care arising out of the new public health. In this view, those at risk of disease, or those who carry forms of disease, are required to take up diagnostic and treatment options and make other adjustments in their lives. Patient expertise also connects with the idea of surveillance medicine and self-regulation. Patient expertise promises a way of engaging with patients outside the confines of the clinic and over the entire life course. In this view, an emphasis on equipping patients to look after themselves is the basis for expanding the effectiveness of clinical care and perhaps in containing the costs of treatment.

Patient expertise is also required because of developments in health technologies themselves. There are many examples of disease where the expertise required to make personal changes presents new opportunities and challenges. The complexities and personal implications of genetic testing are a good example. Some new technologies impact on life expectations, creating opportunities and dilemmas for how to manage the future course of disease and by implication life prospects. An informed, expert patient is therefore required as the location for some of the ethical challenges that arise in treatment. In this view patient expertise may be a rational solution to the growing problem of governing the individual aspects of health and health care via technological development. Expertise is therefore implicated not just in extending the effectiveness of health care but in terms of the ethics of intervention and acceptable health care practice.

Problematising the 'expert patient'

Patient expertise has become a prominent health care strategy in terms of the circulation of informational goods and engagements with new health technologies. But the concept is also open to critique. Wilson (2001) highlights that 'expert' and 'expertise' are problematic terms and can have different meanings depending on context and asks whether a patient can be constructed as an expert, especially when the notion of expertise is governed by structural constraints associated with access to resources (Tang and Anderson, 1999).

Other problems are highlighted by Thorne et al. (2000) and Henwood et al. (2003), where it is suggested that 'governmental enthusiasm' for the initiative does not directly translate into professional behaviour: the notion of the expert patient seems to gloss over entrenched patterns of professional power and lay 'compliance' and professionals can cling to power in their engagements with patients, controlling information and dismissing efforts by patients to theorise or explain their condition (see Shaw and Baker, 2004). In their study of women wanting to learn about HRT, Henwood et al. (2003, and Chapter 5 of this volume) discovered that there were times when GPs 'rejected or dismissed' (605) the knowledge, views and opinions of women who had attempted to become informed about this aspect of their health, particularly if the 'lay expertise' did not complement the medical perspective.

The notion of the expert or informed patient has also been problematised by those who perceive the initiative as restrictive and normative, persuading patients to prioritise medicalised understandings of health and illness (Wilson, 2001; Henwood et al., 2003). Wilson (2001: 139) locates this tension within a Foucauldian framework in which self-management of illness is a manifestation of 'pastoral power', making visible all aspects of a patient's life and self care (see also Gastaldo 1997: 124). This point is reinforced by Thorne et al. (2000) who suggest that, paradoxically, patient expertise both assumes compliance with the medical profession and a degree of self reliance in the management of health.

Patient expertise also necessarily relies on some form of engagement with the science and technology that underpins expertise. Such an engagement creates several challenges for the patient in terms of engaging with technical complexity and the uncertainties built into health science and technology. As health technologies such as advanced treatment methods develop, the knowledge-base of expertise may be distributed in new ways. For example, the interpretation of new diagnostic genetic tests may be difficult for patients. Clinicians may also face a similar problem in that aspects of new technologies may evade their own sense of expertise. Clinicians may increasingly find that they are forced to rely on other knowledge practitioners in health care systems to help them interpret test results for patients. Rapid technological change may therefore reconfigure systems of authority to speak as an expert. Patients also have to take on the provisions and uncertainties that arise in the use of health technologies, something that may not sit comfortably with the personal management of health prospects. In this respect, the focus of patient expertise is not necessarily *informational*. Instead, patient expertise involves the skills and patience needed to engage with the *relational* aspects of the knowledge systems of innovative health technologies. Ironically in this situation, the expert patient may find that their relationship with their clinician takes on deeper importance as a key point of contact with complex and changing health care systems.

The case studies

The case studies that follow provide contrasting perspectives on personal encounters with systems of expertise connected with two innovative health technologies. They raise questions about what counts as expertise and suggest how some new health technologies are innovative because they provide the basis for different expertises. The first case study addresses an ethnographic study of an online 'pro-ana' movement, considering how the Internet is used to help sustain the development of a resistant expertise for surviving with anorexia. The second case study discusses qualitative interviews with people with HIV concerning the advent of effective HIV treatment and the transitions that have arisen in the social relations of expertise.

The case studies together suggest the importance of alternative forms of patient expertise, in particular expertises that resist orthodox constructions of health and illness. They also foreground the uncomfortable way that expertise articulated in policy frameworks connects with questions of citizenship. In particular, health policy predicated on mainstream forms of expertise fails to address the discrediting of people with anorexia or exclusion from HIV treatment on the basis of right to abode in the UK. The case studies also explore how people rework forms of orthodox expertise that presides in the treatment of anorexia and HIV, in an effort to find methods of surviving with serious health issues. In the cases of both anorexia and HIV, innovative health technologies emerge as the means of improving health but also vehicles for experimenting with other forms of more radical, patient expertise.

The pro-ana case study

In the pro-anorexia (or pro-ana) case the ethos of the group facilitates a move towards a 'reflexive consumer,' and the emergence of the pro-ana movement as a significant area for the development of 'new' or 'alternative' expertise. The pro-ana movement promotes a managed approach to anorexia that has sought to redefine it outside medical or other professional discourses. This movement has been facilitated by the development of Internet communication technology, and subsists within a 'shadowy' world of semi-underground chat rooms and websites.

The emergence of pro-ana expertise

In an attempt to explore a construction of the reflexive consumer as an expert and how it impacts on the definition and management of health and illness, fieldwork was carried out in a pro-anorexia online community. It examined the way in which 'consumer autonomy' allows the emergence of an alternative model of health and illness.

'Anagrrl'[1] provides vast information on anorexia. There is a 'potted history' of the pro-ana movement, an interactive area where the users can exchange ideas, provide support and share experiences, achievements and perceived failings. The Anagrrl site attracted participants predominantly from the US, UK, New Zealand and Australia, who posted on a daily basis to the asynchronous message forum. The users were overwhelmingly females between 14 to 42 years, with the majority around the ages of 17 to 20. Most were in full time education, working part time or at weekend jobs to earn extra money.

Anorexia nervosa

Anorexia nervosa is a largely Western and relatively modern condition most often manifested by young females as a desire for fasting and stringent food restriction to achieve a radically slim body shape (Brumberg, 2000). Anorectics, while preoccupied with weight loss and the sculpting of a 'perfect body', often experience overwhelming feelings of anger, fear and isolation in attempting to gain control of a perceived unruly mind and body (Shelley,

1997). The condition moves far beyond an obsession with physical appearance. It is associated with low self esteem, depression and feelings of worthlessness (Shelley, 1997; Lindsay, 2000; Chisholm, 2002). For many sufferers, anorexia becomes a means to retreat from everyday life. It can represent a life style that is predictable and secure, making the processes of food restriction and abstinence difficult to abandon (Chisholm, 2002).

The complexities of anorexia nervosa have received considerable attention from the medical profession and other groups. Several competing explanations have been developed, including biomedical models, where anorexia is related to an underlying organic cause (Urwin et al., 2002, Abraham and Llewellyn-Jones, 1997); psychological models, where the anorexia is seen as resisting the development of adult sexuality (Bruch, 1973); and cultural models, where extreme eating behaviour is a reaction to perceptions of beauty that focus on slim, lean bodies (Gordon, 2000; Grogan, 1999). Feminist models are diverse and see anorexia as a 'metaphor of our times (Orbach, 1993), or a 'crystallisation of culture', where gendered cultural forces inscribe the female body (Bordo, 1993). Others present anorexia as a protest, resisting the complex and contradictory demands of gendered sociocultural forces (Malson, 1998; MacSween, 1993; Gremillion, 2003).

Pro-ana – an exercise in damage limitation?

Anorexia is a complex painful condition, which is socially unacceptable, with eating habits and body shape falling outside the Western definition of 'desirably' slim. As part of the biomedical treatment process, anorectics are often hospitalised, where body shape and food intake are normalised (Abraham and Llewellyn-Jones, 1997). The pro-ana movement is a radical socially-unacceptable approach to the management of anorexia, and has suffered a powerful media backlash, being characterised as encouraging 'normal' and 'healthy' girls and women to adopt anorexia as a glorified diet (Dias, 2003; Doward and Reilly, 2003). Established by those living with anorexia, the pro-ana movement challenges and rejects existing models of anorexia and the normalisation of a healthy body size. This rejection is based on the premise that anorexia often represents the only stability in a damaged life. Maintenance of the disease represents a sense of routine and security.

Becoming pro-ana is a complex and risky process. Many of the girls and women participating in the Anagrrl forum cling to anorexia as a 'macabre comfort'; many do not want to eradicate the illness, but want to find safe ways to live with it. Participation in the pro-ana movement allows the playing out of anorexic routine and ritual in a way that is free from judgement, the threat of treatment and the stripping of a valued lifestyle.

'Resistance' expertise

Dominant biomedical approaches towards anorexia recommend the eradication of the condition and the normalisation of body weight (Abraham and

Llewellyn-Jones, 1997, Luck et al., 2002). For many of the women and girls in the study, maintaining an anorexic state brings greater comfort than eradication. When articulating why they believed in the pro-ana movement, participants such as Lucy expressed the desire to find a place where their eating disorder would be accepted without judgement:

> I believe in the pro-ana movement because it's support for women, girls, men and boys with an ED. It's help understand your ED and learning about it and yourself. It's a place where you can be understood since anorexia isn't really sociably acceptable.

Lucy stresses the way in which she values the web space as a place for learning and communication with others, who provide support for those living with anorexia. As suggested in the following extract, from Angela, pro-ana sites are not concerned to encourage the 'spread' of anorexia, but rather to provide support for those who feel they cannot survive without the condition, until they are ready or decide to choose a recovery option:

> ENCOURAGING ANOREXIA/BULIMIA IS LIKE ENCOURAGING CANCER
> Now you're probably asking, 'Well what the hell is she doing on a pro-ana board?' Simply, I do not view these boards as 'pro-anorexia.' I view them as pro-anorectics. In other words, boards like these, in my opinion, should intend to help out anorectics. Not necessarily to help them recover, just to do whatever possible to put the anorectic in a better state of mind. Some of us are not ready to recover, and if we need food advice, that's what's best for us. Geez ... it's not like we're a team of Anorectics and we should be recruiting people ...

Angela emphasises the way in which the forum offers support that meets the needs of the anorectic living with the condition. It is suggested that many participants feel ambivalent about recovery and want to continue with the routines that anorexia entails. For example, Charlotte indicates that she may want to recover in the future, but presently prefers connecting with other pro-ana supporters:

> Maybe one day I will be 'ready' for recovery but I certainly am not yet – and I am sick and I like to know there are people out there who feel the same way as me.

Susan articulates her feelings on recovery and indicates that recovery would involve a loss of a routine that would be missed:

> I am torn between wanting to recover and having something holding me back from recovery ... What holds me back from recovery – the fear of

losing control of my body, of my life. I'm so used to my 'lifestyle' that I don't want to even imagine going on without the structured schedule of knowing that I WILL work-out everyday and not eat ...

The pro-ana movement rejects the view that 'anorectic eating patterns' can be replaced with 'normal eating patterns'. The participants recognise anorexia as a dangerous condition, but at the same time, acknowledge that anorectic behaviour and feelings provide a sense of security in disturbed lives. Anagrrl provides valuable space for anorectic feelings and behaviour which are socially unacceptable in wider society and attempts to strike a balance: promoting safety in the pursuit of anorectic routines.

There is a profound rejection by those involved in the pro-ana movement of established approaches to managing anorexia, where it is seen as a condition to be eradicated. Through the development of this support group in web space, the pro-ana group have established an alternative, subverted model of anorexia, which not only offers deep and textured insight into the psychology of the state of disease, but could also offer a new and radical method to manage the disease in wider society: a new resistance expertise.

The HIV treatment case study

The HIV epidemic is dynamic, not least of all in terms of the construction of expertise. By the mid 1990s, treatment known as Highly Active Anti-Retroviral Treatment (HAART) became available leading to a dramatic reduction in mortality and morbidity (PHLS, 2000). However, we also need to recognise that many people in the world do not have full access to HIV treatment. For example, many people in Africa cannot afford HIV treatment and many governments cannot afford to fund the required clinical services. The post-HAART era is therefore divided on a global basis. In the affluent West at least, transition to effective treatment highlights the social relations of expertise, in particular, the implications of the uncertain qualities of treatment and treatment-related side effects, contesting the terms of expert power in the management of treatment. But as we will argue, the global division of treatment access has also become a problem for people with HIV living in the UK, highlighting how patient expertise is limited by questions of citizenship.

Uncertainty, side effects and expertise

HIV treatment brings about several questions for expertise to do with inbuilt uncertainties and side effects and by implication life prospects. Expertise on the part of people with HIV can therefore be understood through the idea of reflexive biography engaged with the provisional, and sometimes risk-producing, character of the knowledge systems of HIV treatment (Beck & Beck-Gernsheim, 2002). Clinicians face challenges in relation to the growing technical complexity of treatment and side effects (Rosengarten et al., 2004). Taking HIV treatment implies ongoing self-management of daily routines, diet and use of clinical services for blood tests that help monitor the action of treatment on the virus in the body (e.g. regular tests for viral load and CD4) (Flowers, 2001). HIV treatment also leads to problems of 'secondary risk'. For example, HIV patients have to adhere to their prescriptions as closely as possible to maintain viral suppression and avoid the development of drug resistant forms of HIV which jeopardise future treatment prospects (Sabin et al., 2005). Side effects are also common in HIV treatment. These include nausea, fatigue, mental health problems, the redistribution of body-fats leading to facial and bodily disfigurement, increased risk of diabetes and coronary heart disease. There are also social risks that people with HIV must face. These include managing a stigmatised and sometimes visible identity (as 'HIV positive') as a result of the side effects of HAART, managing the complexity of taking medication across a range of social contexts where positive HIV status may not be disclosed, or managing the risks of potential HIV infection and reinfection with a drug resistant virus with sexual partners. HAART is therefore a technically involved form of medical intervention with implications for patient self-management and treatment.

Contesting expertise

HIV is distinctive because 'expertise' is vigorously challenged. Only since the late 1990s has orthodox medicine been able to effectively treat HIV. The early epidemic therefore gave rise to a treatment advocacy movement that provided a focus for people with HIV to argue for access to experimental treatments and to challenge researchers and funders to develop more effective treatment. This period can be understood as a contest over access to medical science applied to HIV (Epstein, 1996). Treichler referred to this movement as a striving for a 'radical and democratic technoculture' (Treichler, 1999: 280). The key-notes of such advocacy were getting 'drugs into bodies' and 'leaving no stone unturned' in the search for treatment. One achievement of this period was establishing the legitimacy of people affected by the epidemic in the planning and dissemination of treatment research (e.g. clinical trials).

Epstein has also argued that medical power is reasserted in the 'techno-democracy' of HIV treatment advocacy (Epstein, 2000). For example, Epstein discusses the 'expertification' of some people with HIV (Epstein, 2000: web document). People most able to participate in medical systems of expertise (male, white, educated) become 'experts', while those less able to participate are (re)excluded. Erni has also cautioned that the contestation of HIV treatment needs to acknowledge a wider contest about the expertise of medicine. He suggests that the democractisation of treatment technology is limited while medicine in general is oriented to the defence of professional hegemony over expertise (Erni, 1992).

In addition, analysts have identified the duality in the post-HAART situation connected with disciplinary rule (e.g. Adkins, 2002). For example, the promotion of optimum HIV treatment leads to the encouragement of strict adherence to prescriptions and a focus on clinical markers such as viral load, at the expense of more individual concerns such as wellbeing and side effects. This tension in HIV treatment suggests a separation of the expertise of viral management and the expertise of living with HIV. In addition, the sexual and drug using practices of people with HIV have been implicated in the transmission of drug resistant virus, and the question of a possible reduction in the effectiveness of treatment for the individual or population (Sabin et al., 2005). Thus the expansion of treatment options and methods of viral manipulation also mobilise questions of what HIV treatment is for: personal health or control of the epidemic. HIV foregrounds the multiplication of expertises and a contest between the personal management of life with HIV infection and the management of the epidemic spread of HIV, particularly in connection with its post-HAART, drug resistant forms.

HAART therefore raises questions about living with HIV and engaging with the idea of personal expertise. In particular, expertise can be addressed according to two broad themes: exploring how the treating self is achieved in present circumstances; developing an expanded conceptualisation of patient expertise with reference to transition for the post-HAART era.

The next sections (based on Flowers' research) address personal accounts of patients in terms of transition to the post-HAART era; uncertainties, side effects and the limits to expertise; and lastly the (re)emergence of questions of treatment access.

Transition

The following examples show a sense of the changing circumstances for people with HIV and engagements with transitions in medical technology:

> After being diagnosed, one obvious question was the prognosis, in relation to, 'What now?' And they talked about five years in total at that time. This is just pre-combination therapy. Three years quality life with two years diminishing quality.
>
> I think because there are always new drugs coming out I think I've got a good chance as anybody else to have a longer life than I thought I was gonnae have. I mean when considering like ten years ago ... I thought I'd be lucky if I reached the millennium. That was a milestone for me.

These examples depict shifts in the form and scope of life expectations with HIV. They focus on transition. The pre-HAART experience of HIV is one of imminent death. The contemporary situation is depicted as a mixture of calculability and uncertainty, a provisional opening out of the future, engagement with the 'manufactured uncertainty' embedded in treatment, and therefore different questions for the reflexive self (Beck, 1998: 12). The kinds of expertise concerning the management of treatments require an engagement with shifting technology and the articulation of the increasing individual responsibilities of the patient. Once living with HIV meant managing opportunistic infections and malignancies. The present situation seems to comprise living with HIV treatment and therefore managing its positive and negative effects. The orthodox view of transition relies on a sense of a medically determined transfer to a post-HAART epoch (Rosenbrock et al., 2000). But we also need to engage with the social aspects of the procurement and enabling of technological change and the kinds of selves required to make transitions into working treatment.

Uncertainty, side effects and the limits to expertise

The following extract details the dilemmas associated with HAART from an African woman's perspective. We see the consequences of the iatrogenic effects of HAART presented within distinct social contexts:

> I've seen people get better. I mean the pros are more than the cons I think. Speaking from a woman's point of view, I've seen what they've done to some of my friends. You know the way a woman move, a woman is proud of the way she looks, you know a small waist, nice bum, nice legs and suddenly that's all taken away from you. And in a case where you haven't disclosed your status and people start seeing the body changing – that alone is enough to disclose your status. The practical issues around taking medication, how many times a day, you know, the restrictions, are for someone who hasn't disclosed their status who is in a relationship, when do you take your medication? Do you wait for a partner to go to your bathroom? What happens if you go away for a weekend? What happens if you're working for instance and you have to take your medication at lunchtime and your colleagues are there? Someone was telling me about a story where two girls went to ... a party and 8 o'clock is the time when [one of them] should take her evening dose. She took her tablets out and was just about to swallow them, when the other friend said, 'Oh what's that?' She didn't know that the other one was positive. And she said, 'Oh my headache tablets.' 'Oh, I have a headache too. Can I have some?' What do you do then? You know ...

In the following example, the interviewee refers to living long term with HIV treatment:

> I think there is an issue in regards to the long term effects but I think because you know we just don't know anything about it you've just got to travel hopefully ... you have a very hopeful attitude towards the future. But the long term side effects as regards maybe the heart and the lypo

stuff and all the rest of it is something that ... could be right there round the corner you know

This account marks a transition to managing the side-effects of treatment. The self has embarked on a journey, a metaphor that captures how HIV treatment is understood as a biographical watershed mixed with the unfinished story, and in that sense life with treatment unfolds in an uncertain way. The account also reveals the importance of hope as a method of dealing with uncertainty. In these examples, the interviewees refer to self-care as an aspect of a 'war on uncertainty':

I think if you really gave it a good try and watch your lifestyle, you know I think you can delay it quite a bit and I think that's, I think that's what everybody that's got the virus should be concentrating on. To delaying it. Because it's a war of attrition really ...

This extract refers to the idea of self at battle, of collecting and storing up bodily defences against slow attack. HIV treatment is therefore a personal campaign, connoting a patient expert alone in their bodies.

It also seems possible to argue that the transition to effective treatment has mobilised a turn back to fate as a way of understanding the limits of technology. In this example the interviewee describes in fatalistic terms, being excluded from a treatment option:

... big chief of the [HIV Clinic] said he was disappointed that the Trizivir didn't work 'cos he would have liked the Abacavir drug. He thought that was a great drug to be on, but not for me. So that's unfortunate.

In this account, biography is understood as a matter of fate or luck. The sense that treatment options are a matter of the capacities of technology is erased. This account suggests a retreat from engagement with the expertise of HIV treatment into a sense of the life course as subject to chance. Fatalism therefore also marks the limits of the reflexive self. But we can also argue that an acceptance of the limitations of new technology is permitted through a turn back to traditional biographical forms.

Citizenship

For some interviewees, engagements with HIV or treatments were not central questions. For people from African countries residing in the UK, residency was more pressing:

R: I'm not worried about the virus – my worry is whether I will be allowed to remain here in this country.
I: OK – why is that the case?

R: Because I know if I go back I will not have any medication, just they send me back to bed. That's my only worry but otherwise after seeing how people live with the virus here, if I have same care, now I can still live.

The global split in access to HIV treatment was a local reality for people living in the UK. Survival for some was a question of location. Matters of hope and uncertainty were therefore folded inside a more immediate issue of security of immigration status and therefore access to HIV treatment. We can therefore suggest that the global split in access to HIV treatment is not only described by geography. The split emerges in the affluent West as a division in rights of citizenship. In the post-HAART era, the globalisation of the HIV epidemic turns full circle to create a local advocacy agenda concerning treatment access.

Discussion

With reference to the two case studies, we have explored ideas surrounding expertise and we have suggested that expertise emerges in different forms and has varying status, depending on the context of its production. The case studies highlight issues surrounding the articulation of identity and citizenship, where innovative health technologies become vehicles for resisting and defining the limits of expertise.

This can be seen in the pro-ana case study where the group resist medical and social models of anorexia. The pro-ana movement has been enabled by the Internet technology, which supplies a medium for communication and information sharing independent of geographical constraints. Despite efforts to suppress and censor websites, the pro-ana movement establishes a space for participants to 'live' their anorectic feelings and behaviours free from the prejudices of wider society. The role of the Internet in supporting patients has been acknowledged in various studies (Mendelson, 2003), but this study suggests that the Internet is also used by those who reject and resist mainstream models of health and illness and develop their own 'logic' surrounding a condition.

Pro-ana is a lay construction of an illness that is grounded firmly in the experiential and contextual reflections of participants. Anorexia is seen as a disease and 'sanctuary', which provides shelter from disturbed lives and relationships with society.

The pro-ana movement perceives anorexia as a response to social and emotional difficulties, and one that enables individuals to cope. It advocates a 'damage limitation' approach and survival strategy, to reduce the risks associated with extremely low body weight. In their 'resistance expertise' stance, pro ana communities are acknowledging the symbolic importance of anorexia to its sufferers. One respondent summarised this movement as being for 'pro-anorectics' not 'pro-anorexia'. In other words, the movement is there to support its members; helping them manage anorexia safely, without removing the crutch that it provides them.

The pro-ana movement is an example of an emergent community based around resistance to mainstream models of health and illness. The data presented in this case study provides a challenge to medical (and social) definitions of health, illness and embodiment. Rather than becoming 'expert patients' within a medical model, pro-anas have established an alternative, experiential model of anorexia, which provides security for those immersed in the routines and regimens of the illness. Here, expertise is not grounded in medical definitions, but offers an alternative strategy, where a body size that the mainstream would consider unhealthy and morbid is managed in an attempt to sustain life.

The transitions in the treatment of HIV focuses on a different set of technologies, but also offers insight into the nature of expertise as it relates to identity and citizenship. Indeed, there is a post-HAART contest over setting the terms for acceptable personal management of treatment, or expertise. In the West, treatment advocacy has turned into providing access to knowledge about treatments via dissemination projects that digest the vast and growing technical information for patients and carers and span the duality of both the positive (e.g. viral suppression) and negative (e.g. side effects) actions of drugs in bodies. But in the affluent West, there are new questions for expertise that exceed questions of access to knowledge. HAART has brought about a transition from the traditional forms of treatment advocacy that challenged medical expertise in the era of HIV into the construction of individualised engagements with HIV treatment (Flowers, 2001; Flowers et al., unpublished). The treatment experience is therefore characterised by personal engagements with the innovations of HIV treatment with the imperatives of survival (Davis et al., 2002). Reflexive treatment is bent to securing the self in the circulating and proliferating expert knowledges of HIV treatment. But for some, other questions of security, such as citizenship, mean that reflexivity about treatment has less priority (Flowers et al., 2004). We argue that this transition to individualisation creates new questions about treatment advocacy that go beyond access to the substantive aspects of expertise about HIV treatment. Technological innovation of effective HIV treatment emerged in, and was enabled by, treatment advocacy. But in the post-HAART situation, the politics of treatment advocacy is split at a global level. The old agenda of access remains a vital cause for people living with HIV in the developing world.

Transition in HIV treatment can be read as a change in the terms of expertise. It seems that for people living with HIV in the West, there are new frontiers in HIV treatment concerning the personal project of furthering hope, managing uncertainty, personal security and dealing with the capacities and effects of treatments. Our analysis suggests an advocacy agenda for the exterior and interior of the expert systems of HIV treatment. The globalisation of the HIV epidemic has a local dimension where access to HIV treatment remains an issue for people whose residency is in jeopardy. For those more confident about their access to HAART, advocacy needs to address the personal aspects of the expertises needed to survive with HIV treatment. And advocacy needs to move inside treatment itself, to address some of its limitations and dilemmas for patient expertise.

Conclusion

This chapter has outlined the background to contemporary ideas of 'patient expertise'. It has detailed two distinct areas of resistance to medically defined 'patient expertise'. The lived experience, or discourses of expertise, which emerge from both studies discussed within this chapter are grounded within very different understandings of what it means to be an 'expert' at living with a particular health condition. On one level, both reject the simplistic understandings of biomedicine and both present ways of managing the interface of biomedicine with a range of sociocultural issues. Critically the outcome of such subjective expertise is different from that prescribed by biomedicine.

Note

1. The name of the group has been changed to protect the anonymity of the participants.

8
Making Sense of Mediated Information: Empowerment and Dependency

Joe Cullen and Simon Cohn

Introduction

For some time, philosophers of technology have offered us a stark choice between utopia and dystopia. At the pessimistic end of this spectrum, Heidegger's view that modern society is engaged in the transformation of the entire world, ourselves included, into 'standing reserves' – raw materials mobilised in technical processes (Heidegger, 1977), is echoed to a large extent in Habermas' position that the central pathology of modern societies is the colonisation of lifeworld by system (Habermas, 1984). Medicine has frequently been put in the dock as a particular culprit in this 'technicisation' of social relations. As Foucault (1977: 283) put it:

> Medicine, as a general technique of health even more than as a service to the sick or an art of cures, assumes an increasingly important place in the administrative system and the machinery of power.

Foucault's portrayal of a crucial shift from 'practices of the self' (in paganism, Stoicism and in Greco-Roman Aestheticism) to a 'culture of the self' (in Judaeo-Christian culture) through to a present day 'cult of the self' should not be understood simply as a critique of post-modern narcissism, as exemplified by our obsession with 'self-analysis' and 'body image'. It states the case for a more fundamental transition from *epimeleia heautou* – taking care of oneself – to *epeimeleia tonallon* – taking care of others. In the classical period, technologies like cosmology and physics were essentially tools for understanding the position of the self in the world, not the world itself. Other techniques used to support 'practices of the self' included 'literature of the self' – diaries, private notebooks and other forms of *hypomnemata* (personal narratives and individual ethnographies). In the transition towards cult of the self, knowledge and 'techniques of the self' were put to work in the exercise of pastoral power. Particularly from the seventeenth century onwards, this pastoral power underpinned the annexation of knowledge about the body, about health, and about techniques of the self by professionals: surgeons, public health administrators, psychiatrists.

Psychiatry has a particular resonance in this scenario. Influenced by texts like Jaspers's *General Psychopathology* (Jaspers, 1963) psychiatrists, it is argued, mounted a ruthless assault on the social foundations of mental health, further reinforcing the Cartesian separation of 'mind' and body and divorcing mental phenomena from social and cultural factors. This focus on the individual self, supported by highly individuated therapeutic ethics and clinical practices, reflected a determination to replace social, political and ethnographic understandings of madness with a 'technological framework of psychopathology and neuroscience' (Bracken and Thomas, 2001). In turn, the continuing development and refinement of instruments and techniques aimed at the 'medicalisation' and 'technicisation' of the subject – models and manuals of mental disorder; electric shock therapy; serotonin inhibitors; the psychiatric clinic itself – made psychiatry's linkages with social exclusion, incarceration, control and punishment even more evident.

This portrayal of the history of psychiatry, indeed of the history of medicine in general, reflects the uncompromising 'anti-technology' stance adopted by writers like Heidegger, Adorno and, to a lesser extent, Habermas. Yet its dark, pessimistic essentialism – the notion that technology is inexorable, overwhelming and destructive – has been questioned by more recent post-modernists, notably Andrew Feenberg.

Feenberg (1992) has been instrumental in developing two key concepts that are highly relevant to understanding health technologies in the post-modern world: the notion of the 'technical code' and the idea that technologies embody 'civilising choices'. Technical code describes 'those features of technologies that reflect the hegemonic values and beliefs that prevail in the design process'. In conventional interpretations of innovation development, there is an assumption that technologies evolve in a stable, linear fashion. As a technical artefact develops, its essential technical essence becomes refined, embellished and improved through re-engineering and standardisation. This developmental trajectory comes about as a result of the inherent properties of the object itself. However, the reality is that technological inventions are both socially constructed and they also offer a multiplicity of developmental trajectories. For example, the modern 'racing' cycle is typically construed as a direct descendent of the Victorian 'penny farthing'. In fact, racing bikes can be seen as an evolution of a completely different technology – the 'safety bicycle', emerging in Victorian times to provide a less taxing alternative to a penny farthing that was originally bred for speed competitions (Pinch and Bijker, 1984). Although the safety design eventually emerged as the dominant form of bicycle, the 'innovation space' in which bicycles developed in their early gestation was characterised by a number of possible innovation scenarios. This technical ambiguity is known as 'interpretative flexibility'.

More importantly, not only do technological innovations embody interpretative flexibility, they also present 'civilising choices'. In the case of the bicycle, issues of design and development were shaped by a 'contest of meanings' – between the bicycle as a mass transportation tool; the bicycle as a sportsman's toy, and so on. At the genesis of technological innovation, these social constructions and meanings are up for grabs. There is a relatively short moment of 'technological flux' before the technology becomes 'coded' and freeze-framed. It then becomes appropriated into fixed modes of production and consumption. Following stabilisation of the technological innovation, the technical code will determine how the technology is used and embedded in cultural practices. But because the technical code typically expresses the standpoint of dominant social groups at the level of design and engineering, the civilising choices associated with a particular technology can frequently have more profound implications than a simple 'lifestyle' outcome. The evolution of the Spanish flintlock in the seventeenth century, and its adaptation to a colonisation strategy based on genocide rather than the lesser evil of economic domination, is an obvious example. Indeed, Feenberg puts this position more strongly in his assertion that 'the unequal distribution of social influence over technological design contributes to social injustice' (Feenberg, 1991).

It is tempting to envisage the technical code solely in a negative light – as a key mechanism through which technicisation dominates the processes of knowledge creation, and neutralises all forms of counter-discourse and practice (Marcuse, 1966; Gramsci, 1971). Technical coding allows for the subordination of social meaning within the 'technical object', and therefore supports the perception that technicisation itself is the natural order of things. As Foucault has suggested, modern forms of oppression are not so much based on false ideologies as on the multiplicity of technical 'truths' among which the dominant hegemony selects the means to reproduce the system. So long as that act of choice remains hidden, the deterministic image of a technically justified social order is projected (Feenberg, 1992).

Yet it does not always work like this. There are many examples where 'marginal practices' have been brought to bear to change the imprimatur of the technical code; where strong public opinion and participation have altered the course of technological history. In obstetrics, for example, the vocal, and orchestrated demands of large numbers of politicised, and affluent expectant mothers for more 'natural' and less 'medicalised' childbirth practices in the 1970s has led to changes in practices and procedures that today are seen as merely routine (Feenberg, 1991). To take another example, in the early stages of the AIDS pandemic, many commentators argued that the engagement of the medical establishment and the pharmaceutical industry in the 'fight against AIDS' was not always productive or appropriate. Clinicians were generally regarded on the whole as being less knowledgeable and less effective than their clients. As a result, particular social systems and social practices – such as self-help groups, complementary therapy and bereavement management – evolved as major determinants of the medical response to HIV/AIDS, and played a key role in plugging the lacunae in medical technologies and practices. In the UK, from the early 1980s the increasing politicisation of people affected by the virus was instrumental firstly in establishing voluntary-based support organisations and subsequently in securing funding for new and improved technologies and service provision (Cullen, 1998; 2004).

These cases are clear illustrations of the fact that technology is never purely 'rational', but is inevitably embedded in 'action systems' that reflect values. However, the ways in which these 'value embedded action systems' work, and how they relate to the design, development and utilisation of technologies, is not well understood.

In this chapter, we examine two apparently highly contrasting examples of innovative health technologies as live – and ongoing – illustrations of technical coding in action. We focus on the key actors that have influenced the evolution of these technologies; their belief systems and values; how these are embedded in social actions and communicative practices; and how these actions and practices affect both perceptions and realities of illness.

Brain imaging and collaborative knowledge systems

This exploration of 'technical coding in action' combines the results of two projects funded under the 'Innovative Health Technologies' (IHT) Programme. The first focuses on 'The role and effectiveness of collaborative knowledge systems in health promotion and health support'. The second, entitled 'The challenge of recent neurology to conceptions of mental and physical illness', investigates how brain imaging techniques may be constituting a new way in which neuroscience defines normal brain function and specific mental illnesses.

'Collaborative knowledge systems' are part of a family of health technologies, like NHS Direct Online in the UK, that provide digital health information and support services. In our study of such systems, we were interested in a particular type of technology – one that allows both experts and laypeople to work together to produce new knowledge about health. This includes technologies that: allow for evaluation of the information provided; promote interaction between 'producers' and 'consumers' of health information; use 'tacit' knowledge, based on real experience, in health promotion and support; use tools to represent and present knowledge in more intelligible and meaningful ways (for example using video, discussion groups and 'real life' stories); allow different perspectives and narratives about health to be shared to enable a new perspective to emerge. The technical platforms and tools adopted, the target groups addressed, the content developed and the type of information and support services provided vary enormously, but these kinds of health technologies can broadly be divided into four groups.

The first (and largest) type is comprised of systems and services providing basic information. Typical target users are multiple interest groups (diet, sex, beauty, specific conditions like cancer and heart disease). Although most services are web-based, some of these systems use platforms like Digital Interactive TV to reach a wider audience. Health content is consistent with this broad range of multiple interests, and is typically developed by professionals. The 'learning' approach adopted is correspondingly 'transmissive' (conventional top-down expert-consumer), and the strategies, techniques and devices used to promote collaborative knowledge production and interactivity are minimalist (typically links to related websites, database searching and e-mail). An example is 'Ivanhoe', a website providing regular articles on a wide range of health topics.

A second type, encompassing 'web-based public health services', reflects a more focused and less differentiated target user configuration – and hence content repertoire. The main focus is on risk reduction (for example coronary heart disease, sexual health). This type of service exhibits a greater range of strategies and devices used to promote interactivity and collaboration than the first type (incorporating for example content evaluation and editing functions). However, the typical knowledge production and collaboration model adopted is still expert-focused. In this case, collaboration reflects networking between a range of health professionals, whose pooled knowledge and collaborative results are then used as an evolving resource for professionals themselves, and then passed on to 'lay' consumers. An example is 'NetDoctor', a website providing information and support services developed by health professionals, information specialists and patients.

The third type – 'Collaborative Knowledge Constituencies' – reflects arguably the most highly developed type of collaboration system. Essentially the main characteristic of this type is its role in developing communities of practice. In some cases, these communities of practice focus on community-based organisations or common interest groups. '4Woman.Gov' is one such service – a national website funded by the Australian government to promote better women's health through pooling of knowledge and resources. But this tends to be atypical. The typical scenario of use is to support inter-professional working and continuing professional development.

The fourth type – 'Highly-integrated multi-modal systems' – constitutes the smallest proportion of services, and reflects what might be termed the 'emergent generation' of convergence technologies. Convergence in this sense means the integration and configuration of both technical platforms and content. A typical example – the 'Healthology' portal – can perhaps best be described as a 'franchising and branding' web-based service for health information. Basically Healthology is a large evolving knowledge base of health content, created by an extensive network of health specialists. It combines video streaming media with conventional written articles on health issues to provide content that is then 'branded' to suit an individual 'downstream' provider, typically a health website or individual physician. The main aim is therefore to add value to existing digital health content providers by 'enriching' their existing content base.

The technical coding of current neuro-imaging is at a far less advanced stage, but because of this, demonstrates an even greater degree of flexibility. Neuroscience is claiming that its recent developments, first in PET and now MRI, should be regarded as an unparalleled advance in the understanding of the living brain. Rather than simply being concerned with providing images of the organ in all its detail, which first originated in the early 1970s through the use of X-rays and computer tomography so that such structures as tumours can be identified, these new techniques show regions of brain activity through tracking the differential increase in blood flow or the accumulation of tracer chemicals. In other words, the images now can portray the activity of the brain under controlled circumstances, rather than simply its physical form. Thus, though the 1990s was proclaimed as the Decade of the Brain, many now have labelled the first decade of the twenty-first century the Decade of the Mind.

However, perhaps one of the most striking things about these current neuro-imaging technologies is their uncertain relationship with clinical practice. While those techniques that show the structure of the brain have been rapidly adopted into the UK health service, the current scientific excitement over imaging the regional function of the brain has met with greater reservation. In part, this is of course to do with the huge costs of the hardware, the team of experts necessary to conduct the scans – from the procedure through to the data analysis – and simply the length of time needed for the technology to 'trickle down' into routine medical practice. In addition, the fact that many of the processes are not yet fully standardised is likely also to have an influence. Though there are only a few variations of the hardware and software, the current competition to claim the central position is quite fierce. There are different standards for devising the experiments, for the degree of detail that the scans should show (the number of 'slices' that can then be built up into a three dimensional image), the choice of thresholds chosen to determine what activity can be associated to a particular situation, the graphics packages to construct images for publications, and so forth.

There is no doubt that these external factors may be limiting its current uptake beyond the imaging labs. Yet it also seems that the technological shift from structural scans, which could rapidly be assimilated within the existing hospital culture of radiography, to ones of the functioning brain, which effectively require entirely new techniques for interpretation rivalling old diagnostic and talking-based techniques, is proving to cause particular hesitancy. Hesitancy openly exists amongst many of the neuroscientists regarding any specific clinical applications of their work as they continue to develop the technology. Many state that they are simply 'research scientists', trying to understand and explain the living brain, and that therefore they need not be burdened by the possible practical implications of their work. This reflects the simple tactic by which 'research' can be conceived of free of any claims for utility. But in practice, just like much of old neurological work, investigations frequently continue to rely on comparing normal with abnormal people, thereby continually forcing the scientists at least at some personal level to engage with the long-term practical and ethical implications of their work.

This liminal stage of the technology's development lies alongside the more practical and external factors as to why and how it may not yet be fully regarded as a health-related technology. The central point is that this sense of hesitancy is, in fact, a key indicator of it being in a state of flux in which a range of possible influences and actors might determine its eventual trajectory. Thus, though the scientists talk of developments almost solely in terms of hardware – such as more powerful magnets or different mechanisms to detect brain activity – it is clear that the underlying sense that it currently has no definitive application is key to its flexibility. An obvious example of this is the sense of freedom so often portrayed in their experimental practice. Researchers are free to study whatever they want, to design any number of experiments that can be methodologically justified all from the premise that since they have this new extraordinary technology, they have a duty to use it in as many ways and forms as they can devise. Later, no doubt, as coding takes place, the technology itself will become transformed as it shifts from its open-ended status to one that has an explicit sense of its application and utility. In this process, the increasing employment of the imaging technology will be based on previous findings and principles, gradually causing first the experiments to gravitate to a general norm and set of expectations, and then, in turn, causing specific refinements of the technology in line with them.

Currently, the most commonly put forward suggestion for its application is in the process of diagnosis. Continuing the long history in psychiatry of trying to establish a definitive science of mental illness, the prospect that the material basis of many conditions will be established inevitably means that a number of people are already trying to develop the technology so that it can conclusively provide an objective basis for diagnosis. This echoes, to some extent, the 'grand vision' for collaborative knowledge systems. Although they cover, at least in principle, the entire spectrum of health domains, there is an underlying grand conceit that bringing together experts, professionals, patients and citizens will somehow integrate diverse knowledge and experience into a unified and codified evidence base.

The technical coding process

On the surface, the two technologies share few commonalities. PET and MRI systems are examples of what might be called 'high end' health technologies. Collaborative knowledge systems work in a more 'bargain basement' environment. The technology is stable, relatively cheap and, with the exception of what has come to be termed the 'Internet underclass', is more or less a mass audience and mass access set of technologies. This reflects another difference between the two technologies studied. Whereas neuroscience is relatively well-bounded, with a distinctive disciplinary focus and a set of clearly-labelled associated pathologies, like schizophrenia, depression and Alzheimer's disease, the domain of health information and support services is chaotically generic. It encompasses a diverse and eclectic spectrum of people, practices and preoccupations, from tabloid readers curious about IVF treatment, through the 'worried well', to lifestyle obsessives, carers of people with long term illnesses and on to professionals interested in expanding their knowledge base. Another distinctive contrast between the two technologies is that there is a common expectation regarding scanning systems and procedures of direct 'suffering alleviation' (however diversely constructed the definition of 'suffering'). In the case of users of collaborative knowledge systems, no one expects to be 'cured', or at least not directly. What they do expect is to be given keys to various repositories of information that may help them further on down the road towards the resolution of a health issue or condition. This reflects another set of contrasts around knowledge and empowerment. At least superficially, the protocols and practices of brain imaging embody long-established hierarchies and power structures in the medical profession, as well as new and emergent professional roles and practices associated with the integration of the technologies themselves into the clinical environment. In contrast, collaborative knowledge systems are seen in policy and professional circles as a significant opportunity to promote new forms of 'democratised' partnerships between professionals, patients, citizens and others.

However, despite (or maybe because of) these differences, both projects overlap in a number of ways. An important common theme is that they both tell stories about how health is socially constructed. But more significantly, in both cases, the constructions made by those involved in the design, development and use of the technologies on health, concepts of illness, ideas of treatment and appropriate action are all increasingly mediated by and through the technology. In other words, both sets of technologies are 'active' and embody 'value embedded action systems'.

These value and action systems are complex and embody new forms of economic, social, cultural and institutional structures and interactions. A common 'value principle' of the collaborative knowledge systems analysed in our study is a commitment to 'empowering' community groups, the public and patients to influence health services and policies locally and nationally. It is widely believed that establishing a network of diverse members will provide an arena for these stakeholders to develop and share the necessary skills and resources to support community involvement in planning of services and policies. These members include staff who work in the statutory health field, community groups, and individual service users and members of the public who want to get their views heard, together with organizations who provide advocacy, campaigning, education, finance, information, training and other resources to support people to get involved in planning services. Against this background, collaborative knowledge systems, at least on the surface, share a similar mission or vision, articulated through three dimensions: an ideological dimension – common commitment to the development of new knowledge about health issues and problems; a social dimension – the development of new knowledge through collaborative working; an instrumental dimension – the application of the new knowledge generated to enhancing the well-being of patients and citizens.

However, the communicative and discursive spaces in which such systems develop are not institutionally hermetic. They engage with and are shaped by a rapidly evolving and highly turbulent 'external' environment. Health information and support services are increasingly being embedded in the 'digital value chain'. For example, 'travel health' websites offer advice on things like inoculations whilst providing a portal to order goods and services like holidays, insurance, hotels and car hire. Meanwhile Digital Interactive TV (DiTV) services are exploring ways of using health information as a 'hook' to reel in consumers for other services, for example cable/digital TV subscriptions for football channels. Convergence technologies are therefore beginning to promote a similar 'convergence' of digital content that relates health to other post-modern consumer preoccupations, such as 'lifestyle'. Another form of consumerism directly linked to the proliferation of new forms of health information and support systems is online pharmacies. The Internet is opening up spaces for independent pharmaceutical suppliers to circumvent both the legislation governing drug prescriptions and also some of the monopolistic practices perpetrated by the big pharmaceutical cartels. More broadly, the processes that shape the public's 'need to know' in the health sphere are being shaped by complex alliances, for example between government agencies, the media and pharmaceutical companies through processes of 'agenda setting'. As an example, the frequency of public enquiries to the UK government's website run by the National Institute for Clinical Excellence (NICE) is on the one hand highly correlated with the incidence of 'health scares' and health issues that are orchestrated by the media, but is equally responsive to 'issues' raised by apparently 'independent' public health websites that in reality are underwritten by drugs companies. To take another example, in-depth case study analysis of typical examples of collaborative knowledge systems highlighted a process of 'infiltration' and colonisation of knowledge production by commercial interests. This reflects the active involvement in health Discussion Groups of participants whose main task is to act as 'product champions' for consumer goods and services. In one example – a website dedicated to women's health, content analysis of the communications traffic showed that a significant proportion of content was generated by two particular discussees, whose main contribution was to plug the benefits of a particular brand of weight-reduction products.

In the case of collaborative knowledge systems, therefore, the technical code is already showing signs, to use Feenberg's phrase, of the 'unequal distribution of social influence over technological design'. Health professionals, experts and intermediaries play a dominant role in the production, review, dissemination and management of knowledge. Compared with those of experts and professionals, patients and public have a limited say in the social construction of knowledge about health. These findings are at odds with the vision of patient and citizen empowerment that lies at the heart of policies promoting digital health support systems. A common view is that virtual communities of practice work together outside conventional organisational structures and are informally bound together by shared expertise and a passion for joint enterprise. In the cases studied, however, both technology-mediated and non-technology mediated collaboratives exhibited formalised organisational structures, which affected how knowledge was developed and used. Key knowledge creation, management and dissemination functions were carried out by professionals. Public and citizens tended to operate more in spaces inside the system within which they felt more comfortable.

The absence of a patient or citizen role as a catalyst in the development of digital health technologies is mirrored in the governance and management structures associated with their development, evolution and delivery. Typically, governance and management structures reflect conventional, 'top down' models. Steering groups and trustees tend to be from business, academic and health backgrounds. User involvement in shaping the vision of the technologies and services tends to be confined to limited 'feedback' roles.

Similarly, the active system by which a range of cultural values are imparted amongst the brain imaging laboratories of hospitals and university institutions is largely divorced from patient and user groups. There are, however, many different groups of researchers, from a wide range of different disciplines and fields of medicine, who currently see themselves as determining the future clinical application of the technology. Competition is sometimes quite fierce amongst them, as different disciplines and fields of expertise try and establish the direction that implementation of the technology should take. It is their very diversity that currently ensures the technology has not yet been fully standardised. There are consequently very varied technical and procedural ways in which research is conducted, brain images are generated, and divergent ways that the data is analysed and interpreted. Yet, a common feature nevertheless can be glimpsed amongst of all those directly associated with the technology; and that is a set of general values and expectations that propel and motivate the work conducted. These can be summarised as a combined sense of excitement and exploration, and that these remarkable new imaging techniques will allow things to be seen – and hence understood – for the very first time. Some explicitly link this to a revised notion of patient empowerment, by proclaiming that greater understanding of mental disorders will serve to demystify and destigmatise conditions both for individual patients and society more generally. Here, the faith in material explanation, told through images of the living brain, is used to imagine a single plane of biological explanation that places illnesses such as schizophrenia alongside, say, a limb fracture. Thus, one of the key elements in the descriptions by the neuroscientists seen to propel the technology towards a direct application for health is as a 'levelling' device. In effect, that mental illnesses will be one day understood as physical. This powerful motif also echoes the more general assertion made that neuroscience will itself act as a new meta-discipline, embracing such fields as psychiatry and cognitive psychology, to create a more rigorous and scientifically based body of knowledge of the normal and abnormal brain.

What is striking, of course, is that though many patients are central for current research, and though this idea of levelling is widespread amongst the neuroscientists, the patients themselves have little direct role in the emerging system that is embedding values within the practice. The result, however, is that the patients themselves are given remarkable freedom to forge their own, discrete, value system around the technology and in particular the images themselves. Frequently given a printout of initial results of scans as an expression of gratitude, not only will they make their own interpretations and assumptions about what the image represents, but more generally as members of user-groups will use the images to legitimate specific conditions. Initially, this appears to be a parallel example of the same initiative of levelling, since these groups and websites invariably talk about how neuroscience is at least demonstrating that mental illness isn't, paradoxically, 'all in the head'.

Yet, there is a subtle but significant difference in this non-expert emphasis on the biological basis of their suffering. Unlike the scientific researchers, who see biology as the locus for complexity, and who therefore see neuroimaging as a new technique that can at last redress what they regard as crude and over simplistic diagnostic categories and psychological models of mental illness, the majority of patients and user-groups who endorse this technology invest in it the sense that it will redefine illnesses by making them straightforward and incontrovertible. It is, consequently, very separate but superficially shared action values that characterise its current status. This can be seen in the widespread popular representations of the technology in the media and popular culture – as both a technology of extraordinary complexity, and yet also one that can ensure simple clarity and conclusiveness. The tension between these two positions is highly productive, and perhaps serves to encapsulate its current 'undetermined' status.

Outcomes and impacts: how technical coding shapes constructions of health and illness

A central proposition of the technical code is that technologies evolve in discursive rather than reductionist ways. In other words, both brain imaging and collaborative knowledge technologies provide 'benefits' for their users not simply in relation to their innate 'properties' but in terms of how they embody different purposes and beliefs. As we have seen in the preceding discussion, the initial establishment of these two genres of health technologies involves a 'contest of meanings' between different actors and protagonists: surgeons, neuroscientists, schizophrenics, people with depression, concerned citizens, representatives of patients groups, and so on. How this contest is resolved initially will shape the parameters for the ensuing and ongoing process of establishing technological meaning and functionality. Crucially, it will also shape the conditions for how users establish, and extract 'value'.

In the case of collaborative knowledge systems, utilisation patterns are affected by a combination of three main factors: type and profile of the user, their role and function within the system, their level of engagement with the service. Health professionals and other experts are involved at a significantly higher level and rate than patients and public. This pattern is related to: prevailing power structures in the organisation of services, issues associated with the skills, language and technology literacy characteristics of public and patients, and a degree of separation between the organisational 'system' of the service and the 'lifeworld' of patients and citizens.

However, this situation is mediated through the influence of other factors, including: 'health status' and 'loyalty'. People suffering from a health condition use the services at a significantly higher rate than the general public or people who are 'carers' rather than patients. Patients and public who are registered users, and who are collaborators (rather than solely consumers) show higher rates of participation and utilisation of services. The key sociocultural variables affecting utilisation are: literacy, technical competence, socio-economic background, sense of citizenship and community loyalty, and mobility. The key point here is that the efficacy of the systems and services provided – the overall 'utility' of the technologies in providing users with an interpretation of their illness and a set of support functions to help them live with it – will be significantly determined by the degree to which the technologies embody relevant 'values'. This means that 'empowerment' of patients and citizens happens in restricted ways, as a result of the 'colonisation' of key knowledge creation, management and dissemination functions of collaborative knowledge systems by professionals and intermediaries. This reflects the interplay of complex structural dynamics (around the hegemonic nature of professional expertise) and socio-cultural dynamics (around things like the literacy, language and cultural practices of excluded groups). In addition, problems of physical access constraints (especially access to computer terminals) are still an issue, and the evidence suggests that particular groups – notably black and ethnic minority communities – are under-provided for in terms of services.

In terms of outcomes and impacts, therefore, collaborative knowledge systems are associated with some 'liberational' effects. These include: providing access to previously unavailable information systems for particular groups (single parents, people with chronic illnesses); addressing the isolation of patients by creating a 'peer support' space; equipping public and patients with the skills to negotiate therapy strategies with health professionals. The most significant benefits are for people with a health condition. Collaborative knowledge systems and services can support health management and illness coping, reduce feelings and anxieties around 'isolation', and provide inputs to decision-making and therapy choices. There is some evidence that collaborative knowledge systems and services address some of the dynamics that contribute to 'health exclusion'.

For health professionals, the main benefits are associated with: access to state of the art and latest research results and good practices; opportunities for knowledge development through networking; inputs to Continuing Professional Development; improving the effectiveness of client handling (for example through the use of integrated electronic patient records and discussion groups). Ultimately, however, acceptance of these technologies, and perceptions of their value, depends on: the status of the user (particularly professional and social status); the type and severity of illness, and the degree to which users are actively engaged in the construction of knowledge. Moreover, online collaborative knowledge systems do not add value in all circumstances. Health collaboratives using face-to-face interaction and little or no technology also have a key role to play in delivering health policy through group reflection, non-verbal and human mentoring and procedures like action learning sets.

The relatively new status of brain imaging technologies means that it does not yet have such a clear practical value. Health researchers consequently talk of its usefulness in terms of its promise for future application, and therefore regard their current work as merely pre-empting its eventual inclusion into diagnosis and treatment. However, though it is true that much of the most sophisticated and innovative research is yet to be part of regular clinical practice, the influence of these expectations for the future is already having an impact. This then forms the second major set of values that are steering the technology towards a final stable set of applications; the expectations for its future impact are shaping the very expectation and possibilities for the technology in the present. Before any new research is ever conducted, various matters of peer-review and group discussions – necessary because of the expense and vast resource requirements for using the technology – serve to subtly orientate not merely the direction the research takes, but also the assumptions regarding what is appropriate and inappropriate work to be conducted. So it is that neuroscientists are both able to be certain that it will have a major impact in psychiatry and the care of the mentally ill and yet also insist that at present its potential is not yet known and not directly of their concern. Thus, the apparently contradictory effect of its imagined future application already influencing the present is reflected in a similar contradiction between resisting pronouncing in what specific ways patients may benefit from the technology alongside a conviction that it will undoubtedly have a major benefit one day. It is in the spaces that such contradictions open up that values are inevitably – and productively – injected and that coding can operate.

Certainly it is true that the varied use of technologies such as neuro-imaging and collaborative knowledge systems suggests that patients can indeed have an influential role in the eventual coding. In part this might be a story of complexity; not merely of the technical hardware, but also the different necessary networks of expertise and participation that undermine many current developments. But crucially it may also be because at the centre of new technological innovations there are less prescribed relationships between the technology and the user. Thus, the old language of empowerment in public health discourse, which stressed patient 'ownership' as key to revising the professional–patient relationship, may now be being turned on its head. For it is now the distinct lack of any clear ownership – either by the health professionals or the user-patients – that potentially provides the patient with a means to determine and control their engagement.

Conclusions

We have tried to use the two very different research studies to make a number of tentative suggestions. Overall, our argument has been that all too often it is assumed that technology has been established once its material aspects and practicalities have been resolved. But clearly the varied use of that technology, which in turn directly affects how it comes to be applied in the future has a great effect on its continual development. It is thus through the early use of a health technology that much of its sense of purpose is established. We have used the term technical coding to encapsulate the varied ways in which a technology is both interpreted and used. Our point has been to emphasise how coding is not about simply making sense of technology, but about how interpretations are implemented such that the very technology itself becomes established through practice. The case of brain imaging provides a very clear example at an early stage; there are many ambiguities and uncertainties around the technology, largely not in terms of its reliability or validity, but in terms of its use. These hesitancies, nonetheless, are central to its current status; many clinicians beyond neuroscience are highly critical of it, and dismiss its claims, precisely because they have not had a hand in establishing what the technology actually is. Equally, collaborative knowledge systems tend to be viewed more favourably by those professionals who are actively involved in the development and utilisation of health content, yet there is a general perception by 'outsiders' that their value has yet to be proved. Perhaps more significantly, the strongest criticism of these systems comes from 'ordinary' lay people who have an interest in them, but who feel excluded from collaborating more fully in developing new forms of knowledge about health because they feel the knowledge creation space has been colonised by alliances between experts and laypeople with more developed political, social and organisational skills (i.e. the educated, the 'middle classes', and members of pressure groups). In turn, this sense of exclusion leads people to attribute less value to the technology, in terms of its effectiveness in helping them to understand an illness, and in developing 'illness management strategies'.

However, this chapter has illustrated that there are complications to even this description of the uncertainty of the early life of a medical technology. Given the very active role of many patients and user groups, their own facility to interpret and use technologies appears increasingly to have an influence on how they might eventually be encoded. This, then, is a very different form of participation from the original model of empowerment which could be said to be based on old medical expectations of compliance and responsibility. What we glimpse in both cases is how, whilst technology can indeed cause us to regard everything as a resource for exploitation, it too is centrally part of that resource. Ultimately it provides us with new techniques with which to imagine what might be possible. In other words, though technology may allow us to see the world, and ourselves, increasingly as something that we can control and influence, the technology itself is necessarily open to such manipulation as well.

In one sense, the two examples illustrate the emergence of novel forms of social, economic and institutional alliances in the development, design and implementation of new techniques and technologies – between psychiatrists, neuroscientists and surgeons; between citizens, patients and practitioners; between global media conglomerations; public health services; offshore online pharmacies and football clubs. Set against this, the technological design and engineering process still appears to a large extent to reinforce dominant hegemonies, power structures and power relations. Yet both brain imaging technologies and collaborative knowledge systems still offer spaces for what Giddens (1994) has called 'dialogic reflexivity' – the application of technology in performing 'experiments with the self'. In the case of brain imaging, the outputs of a complex technological process and set of techniques – the pictorial representation of the brain itself – allows some individuals both a physical medium to reify for themselves the previously ephemeral phenomena of their medical condition, and also a basis to engage their new understanding of that condition in communicative practices with other social actors. As with collaborative knowledge systems, the technologies, at least in some aspects, enable individuals to participate in developing shared understandings of health and illness in ways that previously were not possible, and which to some extent reach back into classical times by providing people with tools to promote a 'literature of the self'.

Perhaps our argument suggests, therefore, that any critic who sees technology as distinct from the human realm, and potentially ever-encroaching into it, themselves might ignore the very trick of a technology that is fully civilised – that it has embedded into its own normative practices many features of social life. For health technologies, this point is particularly pertinent; we are not simply dealing with external machines, information, communication equipment and systems that regulate and alter our interior bodies, but with technological practices that very soon after their introduction already have many expectations of our bodies and of health determined within them.

Our findings also suggest that policy aimed at putting innovative health technologies to work to promote the decentralisation of health care from acute to primary and self-managed scenarios, with a greater emphasis on health promotion and disease prevention, and the involvement of health practitioners and patients in new partnerships needs to balance and manage three distinctive, and often opposing, dynamics: the integrative dynamic, which is primarily government-led; the hegemonic dynamic, which is essentially market-driven, and the discursive dynamic, which is societal and community-based. In turn, technology design needs to reflect the 'communicative practices' of users and bridge their services and functions directly with the user 'lifeworld', and involve the full spectrum of users in the design, implementation and management functions of systems and services.

Part 3
Innovation, Context and Meaning

9
Time, Place and Settings: Negotiating Birth, Childhood and Death

Jane Seymour, Elizabeth Ettorre, Janet Heaton, Gloria Lankshear, David Mason and Jane Noyes

> These are the shifting sands we have created for ourselves in late modernity as a result of our intrusions into the 'natural' world.
> (Lock, 2001: 191)

Introduction

This chapter examines, through three case studies, how new health technologies are changing the material and social conditions of critical moments of life's transition: birth, childhood development and the end of life. New and emerging health technologies are surrounded by ambivalence and risk: they promise a greater degree of 'human' agency over the bodily processes of birth, childhood illness and death (by enabling 'discrete' surveillance, maintenance of function or increased comfort), but at the same time may be perceived to take on 'a life of their own', placing new constraints around those affected by them (Beck, 1992). These new constraints have essentially social consequences and thus stand in a relationship of perpetual tension with the individual and bodily orientation of the particular health technologies at hand. This tension is, in part, related to the symbolic significance of birth, childhood and death within our culture: a significance which extends far beyond the sum of the material and physical transformations wrought by them (Eder, 1996). Culturally, we tend to place value on representations of birth, childhood or death that can be represented as untrammelled by medical technology: we set great store by 'natural' birth and death, unfettered childhood and the location of the healthy child within a well functioning and relationship-centred family. And yet, paradoxically, we have reached a point in the developed world in which we fundamentally rely on technologies to produce these, especially when they might otherwise be under threat. To this extent, an examination of the interaction between technology and these highly symbolic stages of life, should help to make sense of the risks and benefits wrought by omnipresent biomedical innovations. When applied to moments of life transition, these technologies demand the rupturing of our cultural understandings of the 'natural' – birth, childhood and death, from that which is a 'given', beyond human control and understood in relation to the religious and spiritual realm; to that which is 'made' by humans and understood primarily in relation to the material and secular realm. In so doing, they also transform existing relations in the health care setting, the family and the home.

The first case study reported here looks at the clinical implementation of a computerised decision support system (CDSS) for managing fetal well-being during labour, and thus reducing the risk of infant mortality. The case study suggests that the system, predicated on decision-making as a 'rational' and individual activity, had the potential to introduce new and unforeseen categories of risk, because it involved the imposition of an *individually* oriented system onto complex, partly habituated and essentially diffused decision-making processes which depended crucially on the social structure of the labour ward and its formal and informal networks. The second and third case studies focus on how technology mediates and redefines familial relations. The second case study looks at the domestication of technologies, once only used in the hospital, that sustain life or help to reduce impairment of health. It shows how, while the application of such technologies successfully preserves or improves the quality of lives of chronically ill and severely disabled children, their use is associated with a burden of complex care for each child's family carers; a burden that has far reaching consequences not only for the constitution of 'childhood' and the wider family, but also for the wider social relationships each family is able to maintain beyond the four walls of home. The last case study turns to examine how older adults make sense and understand the application of health technologies during dying, drawing attention to the relationship between end of life technologies and the transformations that are perceived to occur with regard to the natural 'time' to die, the domestic space of home, and family relationships and obligations.

Computerised Decision Support Systems (CDSS) and the management of fetal well-being in labour

The emergence of the concept of the 'risk society' (Beck, 1992) has focused attention on the growth of uncertainties in late modern societies both about the unequivocal benefits of scientific innovation and reliability of previously unquestioned professional judgement (Bury, 1998). Paradoxically, this growing risk-consciousness has gone hand in hand with measurable reductions in the actual risk associated with many activities and is often accompanied by demands for a technological fix for perceived risks.

These features are exemplified by childbirth – an ever more medicalised life event, increasingly focused on the delivery suite of the maternity hospital. The legitimation for medicalisation is the desire to reduce the risks associated with childbirth. Yet the consequence may be a growing expectation that childbirth should be risk free and that the outcome should be a healthy and 'normal' infant. Thus Beck's risk society manifests itself in the delivery suite in the creation of new kinds of risk for patients – that they will have less than optimal care – together with declining confidence in professional judgements. In this context, the development of an increasingly litigious culture is leading to concerns that ever-expanding proportions of the health budget are being absorbed in defending actions and settling claims. This in turn, it is claimed, leads to pressures on health care professionals to engage in a defensive medicine that may generate its own new categories of risk, both for patients and professionals.

Against this background, the research was designed to study the clinical implementation of a CDSS for managing fetal well-being in labour. This initiative was driven by a concern about the level of avoidable infant mortality and morbidity that could be attributed to professional error on the part of midwives and obstetricians. There was a particular concern with apparent difficulties in interpreting the cardiotocogram (CTG), in which fetal condition is inferred from a graphical recording of the fetal heart rate and uterine contractions. The belief at the time was that misinterpretation frequently resulted in unnecessary intervention and, importantly, a failure to intervene when necessary. Research on CDSS was pioneered by an NHS consultant obstetrician who in 1989 founded a research group focused on the development of associated equipment for analysing (umbilical) cord blood gas content – a measure of infant well-being – and in the design of a computerised teaching package designed to improve CTG interpretation skills. (For discussions of the system in the early development phase see: Keith et al., 1994, 1995; Keith and Greene, 1994.)

For social scientists, the system presented a unique research opportunity. First, it would be possible, by studying the implementation phase of the system's deployment, to contribute to an understanding of how innovative heath technologies impact upon existing systems of work. Second, it offered the opportunity to explore the way in which attempts to address one set of risks might, in turn, generate new categories of risk for professionals and patients. These include greater visibility of professional error, deskilling, the challenging or blurring of traditional demarcation boundaries between different professional groups and consequential organisational upheavals including challenges to traditional forms of team-working. Similarly, while the system appeared to offer interesting opportunities for patient empowerment (see Annandale and Clark, 1996: 35; Davis-Floyd, 1994), it was also clear that such a development could make patients more centrally involved in the risk taking associated with the exercise of, hitherto professional, judgement with, as yet, unexplored implications.

There is not space to detail the history of the system's design in all its intricacies. However, our findings suggest a range of complexities that may considerably complicate the implementation of the CDSS and hence mediate its capacity to deliver on design objectives.

Lessons for processes of heath care technology innovation

Despite the widespread recognition that such systems have the potential to improve the quality of care and rationalise clinical decision-making (Hunt et al., 1998; Kaplan, 2001) and despite the continuing proliferation of designs for new systems, few systems have reached everyday clinical use (Kaplan, 2001). The few studies that have explored limited use are highly descriptive and typically presume barriers among nursing staff to new technologies (see e.g. Griffiths et al., 2001). The research sought to go beyond such 'resistance' narratives at the level of the individual and explore the more complex and multiple processes at work.

Design issues

Observation of decision-making in the maternity setting showed that, contrary both to the demands of formal organisational structures and to common sense, much decision-making could not be traced to the exercise of individual judgement and responsibility. Rather, decision-making was a complex social process drawing on formal patterns of supervision and hierarchy, informal modes of communication and consultation, tacit knowledge and experience, and taken for granted 'ways of doing things'.

By contrast, the CDSS design is predicated on the assumption that decisions are taken by individuals exercising rational judgement. Like many modern computerised work support systems, it conceives work as an essentially individual activity, albeit one that can be aggregated with the work of others in the production of a final 'product'. This raises the interesting question of whether the system's individualising characteristic might not pose its own problems.

There is evidence from other settings that a failure to appreciate the significance of the social relations of work often lies behind systems failing to deliver on their design objectives (Mason et al., 2002). Were this disruption to decision-making to occur, it might have implications for a range of issues such as: professional standards and autonomy; processes of professional and occupational socialisation; and traditional apprenticeship models of learning. It is not inconceivable that it could also generate resistance and circumvention (Mason et al., 2002) or be seen as threatening by those required to utilise it. A consequence could be to undermine existing interprofessional co-operation in the case of those labours not subject to continuous CTG monitoring, without substituting a viable alternative source of support for those charged with decision-making.

Thus while, on the one hand, a successfully implemented CDSS might eliminate the scope for obviously 'bad' decisions and 'human error', it might also disrupt the social process of decision-making in as yet unpredictable ways. Most importantly, our results relating to accountability and the social process through which authorisation is secured suggest that CDSS may create quite artificial and potentially counterproductive markers for these. Yet, at the same time, the CDSS is intended to create an audit trail that can identify, in an apparently indisputable manner, who was 'responsible' for each and every intervention (or non-intervention). It is not difficult to see how such a development might lead to new levels of risk aversion that could threaten inter-professional collaboration as well as the exercise of professional judgement.

This is important because an avowed design feature of the system is that it should support decision-making rather than supplant professional judgement. But, matters are not so simple. For example, recent evidence on nurses' use of decision support systems when they are a required tool (as in NHS Direct) suggests operators incorporate the knowledge from the decision algorithms into their own knowledge base, and are often unaware of the influence of the technology on their decisions. They are then able to manipulate the way they use the decision support, to produce recommendations that they perceive to be most appropriate.

It is not inconceivable that similar processes might characterise the everyday use of CDSS. There is clear scope for such manipulation in its design, requiring as it does manual input of key contextualising data on patients. Perhaps the most important conundrum is posed by the CDSS designer's explicit intention that the system should not operate as a substitute for professional judgement. This raises two key problems. First, observation of everyday maternity decision-making suggests that midwives may be more risk averse than doctors – or more inclined to follow set protocols. Were this to lead to a situation in which doctors were more inclined than midwives to disregard the advice of the CDSS, it is not difficult to see how this might generate further strains and challenges. Alternatively, the auditing capability of the system might lead to a reduction in the willingness of doctors to take risks – given the potential for subsequently being held accountable. However, our evidence on decision-making suggests another possibility. Midwives regard themselves as ultimately responsible for their allocated patients, regardless of who makes any particular decision. They exercise strategies for securing their preferred decisions, even from much more senior doctors. At other times, they are resigned to being over-ruled. This complex decision-making matrix suggests that the auditing capacity of the CDSS is likely to be at best imperfect and at worst open to significant manipulation. Unless, it could be guaranteed that each participant in a decision sequence logged her/himself on and off the system, the audit trail would always be partial and often misleading. Evidence both from other technological settings

(Kaplan, 2001b) and from our study suggests that non-compliance is likely to be the norm.

Redefining fetal well-being

As we have seen CDSS was a technology that sought to construct a safe (natural) birth through monitoring the well-being of the unborn infant. In the process, however, it potentially disturbed the social and professional relationships through which the definition of risk and 'well-being' has been conventionally generated in the maternity clinic. Instead well-being is relocated discursively into an algorithm that presumes individualised decision-making, legal accountability and judgements based on markers that have been standardised. This relocates well-being in a clinical space and practice that is, probably, more generative of risk (in the sense of a decline in well-being).

For example the possibility that its apparent success could lead to a demand for continuous monitoring of all labours would inevitably lead to questions about its effects on the quality of the birth experience and on women's capacity to exercise influence over their own care. Consultants and other doctors tended to distance themselves from the 'pregnant body' spatially, physically and mentally and were rarely in the delivery suite except when called in to make decisions or carry out a specific procedure. The opposite was true for midwives who, although risk adverse, had intimate knowledge of patients and upheld their embodied autonomy. For the former, continuous monitoring would not upset their daily temporal rhythm, while for the latter it was envisaged as technological interference in human interactions. In these circumstances we might expect that midwives would find ways to manipulate a system that had this effect. Indeed there was some evidence from early prototype trialling that this would, indeed, happen.

Medical technologies and the time of the body: the case of technology-dependent children

'Technology-dependent children' is the term used to describe children who use medical technologies to compensate for the partial failure or loss of a vital body function, and who require technically skilled carers to look after them (US Congress OTA, 1987). Examples of such technologies include renal dialysis, assisted ventilation, artificial nutrition, intravenous drug therapies and oxygen therapy. Advances in these and other medical technologies have enabled more children in the UK and other developed countries to survive and be cared for at home rather than in hospital. In 2001 it was estimated that there were up to 6,000 technology-dependent children living in the community in the UK (Glendinning et al., 2001).

The present study was designed to examine the experiences of technology-dependent children and their families living at home in the north of

England. Informed by sociological studies on time and, in particular, work exploring the multiple temporalities that are characteristic of social life (Adam, 1994, 1995; Daly, 1996; Giddens, 1984; Young, 1988; Young and Schuller, 1988), the study explored the temporal organisation and time consequences of the care regimes for the children and their families. Here we examine the performance timescales of medical technologies and consider how these relate to the 'natural' time of the body and the 'social' time of the children and their families. The analysis is based on data from interviews carried out with 46 parents, 13 technology-dependent children and 15 siblings from a sample of 36 families. There were a total of 38 technology-dependent children in the sample, all of whom were using one or more medical technologies on a daily basis or had been until recently. More detailed information on the aims, methods and findings of the study is available elsewhere (Heaton et al., 2003a & b; 2005a & b).

'Natural' and 'artificial' body time

In technology-dependent children, some of the functions of their bodies are performed artificially, with the aid of medical devices, while others are carried out naturally. Without the technology, the children may not survive or their health status and quality of life would be more seriously impaired. However, being dependent on medical technologies can place considerable time demands on the children and their families who provide most of the care they require when the children are living at home. Part of the time demands relate to the time of the technology itself – how long it takes to perform its functions – and part relate to the temporal organisation of the care regimes that are associated with the technology; that is, how the use of devices and the provision of care are managed in the context of the manifold rhythms and routines of everyday life.

For example, different medical technologies vary in the extent to which they duplicate the 'natural' or 'normal' rhythms of the body and/or have to be regulated by users and carers during the day and/or at night. Some devices, such as pacemakers (which were not featured in the study), more or less restore the natural rhythms of the body, work automatically and are a relatively non-cumbersome appendage to the user's body. Other technologies, such as renal dialysis and artificial nutrition, used by several children in our study, artificially perform functions of the body in different timescales to that performed normally and in ways that have to be carefully managed by users and/or carers who have been trained to operate the device(s). Users' and carers' movement in time and space may also be limited by the portability and number of devices used, by the amount of time users spend physically 'plugged in' to a machine, and by the ease of use of devices while on the move or when in different settings outside the home and hospital.

The performance timescales of two of the medical technologies used by the children in the sample – renal dialysis and artificial nutrition – are briefly described below to give an indication of what they involve. All eight children on renal dialysis received continuous cyclic peritoneal dialysis (CCPD) at home. In this, and other types of dialysis, the precise timescales of treatment varies depending on what has been prescribed for the individual child, but there are nevertheless differences in the performance timescales across the different types of dialysis. For the children in our study, CCPD lasted for 9–10 hours a night for six or seven nights a week. The CCPD regime differed from that which some of the children had previously experienced when on other forms of dialysis, namely haemodialysis (HD), which was provided at hospital on three days a week over three hours, and/or continuous ambulatory peritoneal dialysis (CAPD), where fluid constantly dwelled in the body and was exchanged in three to four bag changes during the day, each taking around 30 minutes. In contrast, the 22 children receiving artificial nutrition experienced a more variable regime depending on factors such as their age and size, the method of tube-feeding (via a gastrosomy or naso-gastric tube) and whether they were totally dependent on artificial nutrition or could also feed normally. Thus, at one end of the continuum, some children were fed by tube at two-hourly intervals during the day and/or continuously overnight for 10–12 hours a night for between five and seven nights a week. At the other end, one child simply had a short supplementary feed while asleep.

It should be emphasised that several children were using two or more technologies (and/or on complex medical regimens) that added to the complexity and overall time demands of their individual care regimes. Furthermore, as well as overseeing the use of the technology, carers of technology-dependent children (usually parents and especially mothers) also spent time on related technical tasks, including preparing the equipment and managing supplies. While 12 technology-dependent children received assistance from trained carers when they were at school, parents mainly looked after the children at other times. Thus, the daily rhythms and routines of caring for a technology-dependent child varied from family to family depending on the number and type of technology used, the performance time of the technology, the time demands of related technical care tasks, and factors such as the availability of trained carers within and without the family to support the children and their parents and siblings. In the next section we look at the extent to which the families were able to fit the time demands of the care regimes into other domestic, school, work and social schedules.

Tensions in technological time and social time

While families were, to varying degrees, able to incorporate aspects of the care regimes into their everyday routines, at the same time, they experienced a range of difficulties as a result of the time demands of the care regimes being incompatible with other domestic, institutional and social schedules.

For example, both dialysis machines and feeding pumps were often used overnight, enabling the children to receive the requisite number of cycles/volume of food while they were asleep. However, parents from 22 of the 36 families reported suffering regular sleep disruption (getting up at least two nights a week) as a result of having to get up to tend to the machine (for example, when it alarmed because of tube kinks, blockages or leaks) and/or to look after the child (for example, when s/he needed turning in bed or disconnecting from the machine). Parents claimed to be regularly getting up between one and ten times a night, and more frequently when the children were unwell. One parent stayed awake five nights a week (overnight care was provided on the other two nights) to monitor her child for when his airway needed suctioning. While it was parents who mainly reported experiencing sleep disruption (some parents observing that their children generally slept through the alarms), a few children also said their sleep was disturbed and/or that they felt tired through having to get up early as part of their regime (especially on school days).

The time demands of the care regimes also limited the extent to which the children and their families could participate in education, employment and social life in general. For example, in addition to missing school when unwell, some children also missed school because their medical appointments were scheduled during school time. Some of the children's siblings had also missed school or college because of the needs of the technology-dependent child; they also helped out by taking on extra domestic duties and/or care responsibilities themselves that limited their social lives. Families also experienced some difficulties combining working and caring. Twenty-one of the 30 fathers in the sample were in paid work compared to just eight of the 35 mothers. Eight of the fathers worked shifts and two had their own businesses; all but one of the working mothers had part-time employment. These patterns of work generally suited the families who combined working and caring. However, several fathers had given up work or changed their jobs and hours to better fit with the needs of the technology-dependent child and/or their siblings; some mothers also felt unable to work because of tiredness and/or the lack of flexible employment that would fit with the children's care regimes and allow time off when the children were unwell. Parent's social schedules were also fitted into and adapted around the children's care regimes. For example, respite care and babysitters were mainly used on those evenings when the children were not using their devices as the families found it difficult to access technically competent carers on nights when the devices were being used.

Time, technology and social inclusion/exclusion

While advances in medical technologies have enabled more technology-dependent children to survive and live more normal lives at home with their families, at the same time, the time demands of the associated care regimes create some difficulties for these families. This brief analysis of the timescales of medical technologies has shown the ways in which the 'natural' time of the technology-dependent body and the 'social' time of users and carers are configured by the technologies and associated care regimes. It has also highlighted some of the tensions and dissonance that exist between the rhythms and routines of the technologies and care regimes on the one hand, and various domestic, institutional and social schedules on the other, and shown how this contributes to the social exclusion of the children and their families.

The findings of the study have implications for policy concerning the organisation of services and for the future design of devices. Changes and developments that would help to minimise, ease and offset the time demands of using and overseeing the use of medical technologies include the following: increased respite provision for the whole family so that parents, children and siblings can have a break together and enjoy a good night's sleep while paid carers look after the technology-dependent child; better co-ordination of medical appointments so that technology-dependent children do not miss school unnecessarily; more flexible employment opportunities for parents of technology-dependent children; more accessible settings where devices can be more easily used outside of the home and hospital; and improvements in the design of devices so that, for example, the incidence of alarms at night caused by tubes kinking, becoming blocked or disconnected, is reduced thereby helping families who suffer regular sleep disruption at night partly for this reason. Such initiatives would potentially help technology-dependent children and their families to more easily combine using and overseeing the use of medical technologies with participation in education, employment and social life in general.

Technology and natural death: a study of older people

Medical technologies instituted since the Second World War in the West have made a diagnosis of dying difficult and the management of the transition from living to dying fraught with ethical and legal problems, giving rise to unforeseen iatrogenic complications (Illich, 1976; Ellershaw and Ward, 2003). Cultural views about what constitutes 'natural' or 'good' death (Sandman, 2004) what the meaning is of suffering, and divergent stances about the value of autonomy, have strongly influenced debates internationally about how health technologies should be applied during the dying process. This draws attention to the way in which death is both a socially organised and a profoundly physical transition (Turner, 1996: 198). These themes are visible in the deliberations surrounding the reviews of legislation regarding mental capacity and assisted dying in the UK and in media and professional and public discussions of these matters.

In the developed world, death is now most likely to occur at the end of a long life but the co-morbidities of old age mean that the diagnosis of dying in older people is often only made by exclusion; many older people facing death are therefore likely to have few opportunities to reflect on end of life care options or how these fit their values and circumstances. Moreover, ageist stereotypes, only now coming under sustained critique, mean that many older adults with chronic life limiting illness face fundamental inequalities in health and social care provision. For these older adults, the last months and years of life involve 'living on thin ice' (Lynn and Adamson, 2003) physically, socially and economically. There is little or no knowledge of older people's perceptions about this stage of their life or their practical, social, spiritual and existential concerns. Yet there are high levels of anxiety among health professionals about the risks, benefits and consequences that flow from attempts to share end of life decisions with patients and their family carers, a practice that is universally recommended within professional guidance (BMA, 2001; GMC, 2002).

This project used an innovative methodology developed in partnership with community groups to explore these issues with older people living in Sheffield, UK. We explored views about two categories of technology used in end of life care: 'life prolonging' treatment and 'basic care'. The first category refers to 'all treatments which have the potential to postpone the patient's death' (BMA, 2001), including cardiopulmonary resuscitation, chemotherapy and artificial nutrition and hydration. The second category refers to 'those procedures essential to keep an individual comfortable' (BMA, 2001). This includes the setting in which care is provided, pain relief and the management of distressing symptoms as well as nursing care such as the provision of warmth, hygiene and the offer of oral hydration and nutrition. We also explored the linked issue of decision-making: first, by examining perceptions about how critical decisions about life prolonging and basic care technologies should be made in *present* time, and second, views about preparing for *future* incapacity through the device of the advance care statement.

Methods

Seventy-seven older people from three age cohorts (65–74, 75–84, 85 years and over) and from three socio-economically contrasting areas of Sheffield, UK took part in interviews, focus groups, and a discussion day at the end of the project. Pictures, story boards and media extracts were used to enable discussion during data collection. The research team was assisted by an advisory group which included research participants.

Life prolonging technologies: a time to live and a time to die

    Hilary: I've had a cancer operation and if I'd have been meant to go then I would have gone and yes, I think you've a time to live and a time to die.

Harry: Book of St Mark's –
Hilda: But then reverse of that would to be something like I had, an infection, if I'd stayed at home probably an hour or more – that would have been it, I get to the right place at the right time and against all predictions I still made a brilliant recovery.
Hilary: Yes, well you weren't meant to go was you? No – else you'd have gone.
Hilda: That was thanks to science though, without a doubt.
(extract from focus group discussion)

The timing of death was a central aspect of respondents' deliberations in this study. Life prolonging technologies (we looked at resuscitation and artificial feeding) were evaluated primarily according to their potential for disrupting or supporting the 'right' time to die. In expressing thoughts about issues on which many clearly had never had an opportunity to debate, critical death-related experiences were frequently recounted in the context of predominant religious and cultural images of death and of medicine. While references to religious beliefs were important for many, others took a more secular stance, while a third group integrated medical and religious discourse with faith in both communicated as an important aspect of facing inevitable mortality and illness in later life. Linked to ideas about timing, were views about the risk and futility of life prolonging technologies sustaining 'a body'. This was seen to constitute a denial of essential humanity and an interruption of the 'naturalness' of the dying process. Respondents recognised that the application of technological innovations to the management of dying had transformed a social order of dying in which 'doctor knows best' to one in which patients, clinicians and their families were caught in a shared dilemma. It was recognised that, in these new circumstances, families had to be ready to assume a degree of responsibility for representing their dying relative and that new difficulties were associated with this. These were revealed most clearly in respondents' deliberations about the risks and benefits of making an advance statement (Seymour et al., 2004) as this exchange about artificial feeding in one of the focus group discussions shows:

Frank: See, and if it happened to me I wouldn't want my family to endure seeing me there like that, I would want the [feeding] tube removed.
Emma: Oh, if it were *me* it would be a different story, if, I would want 'em you know to do that, but I don't think I could do it to my child.
Frank: I'd want to get it over and done with so that they wouldn't have the hassle of trying to make minds up, trying to make their minds up. In fact I would be prepared to stipulate that now if I was making a [living] will.
(extract from focus group)

Basic care technologies: making the passing as easy as possible

Catherine: That they can make life, or the passing as comfortable and as, as easy as possible. I don't believe in euthanasia but I believe in making the passing as easy as possible.
(extract from focus group discussion, reported in Seymour et al., 2002)

In considering basic care technologies (we looked at pain and symptom relief, nursing care and location of care) the terms 'comfort' and 'love' were used by respondents to describe good care during dying. An idealised 'good' death was that in which pain relief and, where necessary, palliative sedation serve to provide dying people with an easy, comfortable and quiet death. In this, bodily symptoms of distress are discreetly controlled and death occurs at an appropriate time and place. Experiences of bereavement were drawn upon by many participants to describe how they understood the role of pain relief and sedation during dying. In these, the distinction was made by many between 'making the passing as easy as possible' and hastening death by euthanasia.

Caring for some one who is dying demands close attention to their bodily needs through nursing care. Hallam et al. (1999) argue that ways in which attention is given to the body influences the recipient's sense of personhood or, in the case of someone who is imminently dying, that which their close companions perceive on their behalf. In the present study, some settings of care (particularly care homes and hospitals) were perceived to be associated with bodily care practices that undermined the personhood of dying people and therefore threatened the quality of death that could be achieved. Illness and disability were seen to compound the influence of the 'mask of ageing' (Featherstone and Hepworth, 1991) perceived by professional care staff, and it was feared that this meant that one would be regarded as a 'non person' or a 'grey person'. Some couched this is terms of being viewed as 'just a body', with one's needs for respect and human regard as a person at risk of marginalisation.

The private home was perceived by most as the ideal place in which to be cared for at the end of life, because of its symbolic meanings of love, independence, familiarity and as a repository of memories. However, a range of practical and moral problems associated with care at home were recognised that similarly threatened personhood: fears of dying alone; worries about being a 'burden' to family; and concerns about the caring skills of family carers and the risk of receiving inadequate symptom relief. The following extract, from the daughter of a woman who died with dementia, describes such fears:

Fay: When mum was in such a bad state near the end and it was obvious she was going to die within days and I was just more or less left to see to her. Now I was panicking because I'm thinking what does she

need? You know I didn't even know how to put the pillows to make her as comfortable as a nurse and the district nurse could see that and it was her who said 'I think you need intensive nursing now don't you?'
(extract from focus group; reported in Gott et al., 2004)

However, for some the presence of professional carers in the home, and the need to accommodate technologies to aid caring, were seen to compromise the public /private boundary and transform 'home' into a form of institutional care setting.

The ideas about technologies applied in end of life care that are explored here can be understood as dynamically created, with individuals interacting 'with technologies in ways that texture their self hood and ways of thinking about and using their bodies' (Lupton and Seymour, 2000: 1,860). They shed light on how perceptions of the use of life prolonging technologies during dying are attended by a complex mixture of beliefs about care, personhood, family and obligation. Participants' understandings of 'natural death' depended not on the simple presence or absence of particular types of life prolonging or basic care technologies, but on the extent to which any technologies employed during end of life care might be applied in ways that reflect and allow space for moral concerns to protect, care and represent the person who is dying.

Conclusion

Ettorre (2000: 416) observes how the construction of the body through reproductive technologies, plays down the importance of human agency and at times, emotional labour. In the examples presented here, what are revealed are the essentially human struggles of persons as they work to incorporate now ubiquitous technologies, and mould them into their own lives. These are struggles that are both moral and social, involving the negotiation of private and working relationships, the reorganisation and reconceptualisation of time and domestic space, all against a backdrop of attempts to properly fulfil professional and familial obligations.

Taken together, the three case studies reveal how persons confronted with innovative health technologies have to make decisions about how these fit with their preferences as well as their understandings of their wider roles and responsibilities within their social world. In case study one, there was resistance to the IHTs in question from those expected to apply it. This is explained in terms of the design of the IHT being based on an individualised rather than a collective mode of decision-making. It is clear that the outcomes of the IHT had a differential influence on the various professional groups in the work setting; being temporally 'out of step' with the work rhythms of the midwives, but having comparatively little impact apparently on the work of doctors who had a more episodic presence. This arguably reinforced rather than alleviated traditional power divisions that are known to constrain team working and innovation. How the IHT would influence the women to whom it was applied is less clear: but arguably this negative consequence would have unintended implications for the quality of care of they received, and for the range of choices that were possible during labour. In case study two, there were similar tensions in the temporal design and operation of the devices and the preferred temporal organisation of social and family life. Failure to understand these rhythms and the social ordering on which they are based is likely to lead to a technology out of time and place, as in the first case study. In case study three the way in which IHTs were perceived to prolong life and facilitate death presented older adults with new choices and dilemmas. They understood these only in the context of wider and more established social and moral considerations.

What conclusion can we draw from our studies of these fascinating areas? The introduction and use of technologies are negotiated processes informed by existing social structures and relationships, and need to be seen in the context of fundamental moral values. Furthermore, the injection of innovative health technologies into cultural encounters and social relationships impacts on the aesthetics of the one's life experience. Similar to other medical technologies (Rose, 2004: 121), an ethics is engineered into the make up of these technologies as they embody and incite particular forms of experience. This raises important issues around the discourse of embodiment and the unique interplay between embodied selves, the ethics of experience, 'nature', human biology and machines. The design and evaluation of IHTs in these fields must take account of these social and moral dimensions if they are to achieve the primary outcomes envisaged from their application and avoid far-reaching disruption to social relationships and sensibilities.

Acknowledgements

Thanks to colleagues involved as co-researchers in the projects reported here: Sam Ahmedzai, Gary Bellamy, David Clark and Merryn Gott; Patricia Sloper and Robina Shah.

10
Replacing Hips and Lenses: Surgery, Industry and Innovation in Post-War Britain

J. S. Metcalfe and John Pickstone

This chapter brings together two case studies of medical innovations, and two authors from different disciplinary backgrounds, albeit from the same University. The innovations in question have proved hugely important in improving the quality of life for increasingly elderly populations. They have many similarities and interesting differences, and that balance provides a good opportunity for comparative analysis, drawing here on the approaches to medical innovation which are characteristic of economists interested in innovation (see Metcalfe, 1998) and of historians focused on modern medicine and its technologies (see Pickstone, 1992; 2000a).

The innovations in question are the intraocular lens (IOL) for the restoration of vision in patients with cataracts (studied by the project in the Centre for Research in Innovation and Competition, and the total hip replacement (THR; also known as total hip arthroplasty) for the restoration of movement and reduction of pain in patients with arthritic and related hip disease (studied by the project in the Centre for the History of Science, Technology and Medicine). In both cases, Britain after World War II was a key site of innovation and particular surgeon-inventors have been credited for the successful development of the prostheses. For the IOL, it was Harold Riley, senior eye surgeon at London's major eye hospital, the Moorfields. For the THR, it was John Charnley, an orthopaedic surgeon based in Manchester who created a research and development unit in a former TB hospital at nearby Wrightington. Both gained knighthoods, and Charnley also became an FRS.

In both cases, however, many other surgeons were involved and there was widespread and prolonged experimentation with a range of related techniques. The surgeon-inventors initially worked with small manufacturing companies, but in both cases now, the manufacture, and to a considerable extent, the technical development of the field, is in the hands of large multinational corporations, such as Allergan and Alcon, in the case of the IOL; and Depuy, Zimmer, Stryker/Howmedica, Biomet, and Smith and Nephew for hips.

The two devices have radically altered specialist surgery and its services, and have helped create new industries which in turn have impacted on the professions and services in ways which bear comparison with 'big pharma'. Both cases raise many questions about professional and industrial attitudes and interactions, about designs and safety, and about the assessment of long-term success, including the regulatory roles of government.

In what follows, we first give brief narratives of the IOL and the THR. We then systematically compare the two cases under the following headings:

- The design spaces and medical contexts
- Innovation systems
- Surgical innovation
- Professions, industry and government regulation.

Of course, no narratives are innocent. The summary histories we present are designed to set the scene for the subsequent analyses, which themselves reflect our different approaches as scholars.

The IOL story

Cataracts – the progressive clouding of the eye's crystalline lens – are the most frequent cause of defective vision in later life. 'Couching' cataracts (pushing the clouded lens out of the line of sight) was a traditional operation practised in the eighteenth century by itinerant 'oculists'; and eye surgery was one of the first fields of specialism as medicine took its modern form in the nineteenth century. Before WWII the standard surgical 'cure' was removal of the lens; a successful operation meant that light could now pass to the retina but without being focused as a clear image (the condition known as aphakia). The only corrective method was to use 'pebble glasses' – which, at best, provided poor, distorted, post-operative vision (Kaufman, 1980). The risk of infection and of collateral operative damage meant that this was a procedure of the last resort for most patients (Linebarger et al., 1999).

In the first half of the twentieth century the dominant operative method was the removal of the entire lens *within* its capsular bag. But Ridley had formed the view that this method (ICCE) was inferior to the alternative method of extra capsular cataract extraction (ECCE) in which the lens capsule was left *in situ*. Crucially, he believed that the prevailing lens extraction procedure was 'but halfway to a cure which is complete only when the lost portion is replaced' (Ridley, 1951: 617; 1952, 1958). The other half was to find a way to put in a new lens, and here Ridley may have been influenced by his knowledge that plastic contact lenses, invented in the 1930s, seemed to be tolerated. Ridley's invention also depended partly on his experience in WWII, when injuries to pilots had indicated that Perspex, like glass 'shrapnel' would lie 'inert' in the eye.

Here was an ideal material, inert, and light (almost the same specific density as eye fluid), but industrial Perspex contained too many impurities. Lacking the requisite chemical and engineering knowledge, Ridley joined forces with John Pike of Rayner, a small ophthalmic company in London, and John Holt of ICI (the giant British chemical company) to develop 'Perspex CQ' (Clinical Quality). Ridley, Pike and Holt worked in secret on the use of ECCE technique, the design of the rigid lens, its manufacture from Perspex, and the insertion of the lens in the posterior chamber of the eye. They collaborated on a non-commercial basis (for fear of the wrath of fellow clinicians), and Rayner agreed to manufacture the lens and supply them at cost. Here we find the first tentative shaping of a local distributed innovation process, bringing together the complementary capabilities of the clinician, the technician and the industrial chemist.

Ridley implanted the first IOL in a 45 year-old woman on 29 November 1949. The operation took place at St Thomas's Hospital, London, using the ECCE technique (the removal of the cataract predated the insertion of the implant by three months). Ever aware of the radical nature of his multi-dimensional invention, Ridley delayed until July 1951 before he announced his results to the annual conference at Oxford, the premier meeting of the British ophthalmic community (Hamilton, 2000). By then, seventeen months

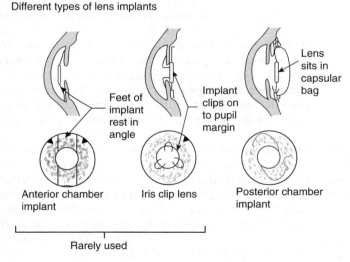

Figure 10.1: The eye as a design space

after his first procedure, he had implanted eight Rayner manufactured lenses. Not unexpectedly, the response at Oxford was largely hostile.

In the early years of the IOL, the first lenses were too thick and heavy. Furthermore, they were turned by hand and consequently varied from copy to copy. Until 1957, no satisfactory method existed for sterilising the lens, so post-operative inflammation was a major problem; and so was dislocation – the slippage of the implant out of the line of sight. As a result of these complications, perhaps 15 per cent of patients subsequently had to have their lenses removed, a failure rate that would certainly colour the view of other surgeons against the technique. Ridley, however, held firm to the value of his methods, maintaining that they gave the patient a better visual field, an optically normal eye, and avoided the problems of binocular rivalry.

Though many surgeons shunned the operation, Ridley attracted a following, who eventually formed the International Intra-Ocular Club. Meeting first in London in 1966, it subsequently became the European Society of Cataract and Refractive Surgeons – which now contains over 2,500 members in 100 countries and an eponymous journal. Within the early community, many other surgeons soon emerged as inventors-cum-innovators, seeking to improve on Ridley's design.

The complications, and the demanding nature of Ridley's technique, encouraged other surgeons to experiment with lenses placed in the anterior chamber, which is technically easier. The pioneer in 1952 was Baron; but Strampelli, working in Rome, implanted the first widely accepted anterior chamber lens in 1953, followed by the eminent Barcelona-based surgeon, Barraquer. But by the time Barraquer presented his results at the Oxford Conferences of 1956 and 1959, it was clear that the anterior chamber lens was creating new design problems. No accurate method existed for measuring the magnitude of the anterior chamber, so the rigid lens might touch and irritate the inner surface of the cornea, the endothelium. Attempted solutions included lenses with more flexible haptics, and lenses with open or closed nylon loops to lessen the irritation to the angle of the chamber (Dannheim and Barraquer designs). Choyce worked with Raynor to improve the Strampelli lens (Choyce, 1960), and two of his lens designs (the M8 and M9) were the first IOLs to gain approval from the FDA in the early 1970s.

The other major development in design was the introduction of iris-supported lenses; the first, in 1953, was Epstein's 'collar-stud' lens, followed by Binkhorst's 'iris-clip' lens in 1958. (Binkhorst, 1959: 573–4). By 1959 Binkhorst had carried out nineteen implants; but despite the initial promise of this design, long-term, multiple complications in relation to the stability of the lens and iris reactions led to their eventual abandonment (see Apple, 1984: 10).

Clearly, Ridley's invention and innovation had stimulated a great deal of creative, experimental endeavour to contest and explore the design space of the eye. Yet, despite this long sequence of inventions, the method remained a problematic operation limited by the supply of super-skilled surgeons.

Of all the developments that have transformed Ridley's innovation and operative method into a mass procedure, by far the most important was the adoption of phakoemulsification techniques for cataract extraction – developed by Charles Kelman, a Professor of Clinical Ophthalmology in New York, who had established his credentials as an inventor in the 1960s with a sophisticated cryoprobe for the removal of cataracts. In 1963 he turned his attention to reducing the size of the incision in the eye. Rotating mechanical cutting devices proved fruitless for the hard cataracts found in old people (the soft cataracts occasionally found in younger people were easier to break up and remove by aspiration); but Kelman chanced upon a possible solution in an ultrasound device, and experimented for many years using the high frequency energy of a vibrating needle to fragment a cataract, which could then be sucked clear of the eye through a much smaller incision than that traditionally associated with the ECCE technique.

Improvements followed quickly and the first crude machines were made available commercially in 1970, signalling the shift of commercial cataract innovation towards the USA. The device was patented in collaboration with an engineer, Anton Banko, and consisted of the ultrasound needle, a supply of irrigating fluid, a pump to evacuate the debris from the liquefied cataract, and a control mechanism for the surgeon (Kelman 1973; 1991). But, obviously, it is pointless to make a small incision to remove the cataract with the phako technique, if one then has to make a larger incision to insert a conventional, rigid or semi-rigid PMMA lens; so phako stimulated the development of new kinds of foldable lenses. But crucially in this case, the new knowledge lay beyond the ken of clinicians.

In the latest stage of IOL development, lenses are 'injected' into the eye and unfold within the capsular bag – an innovation which can be said to complete the revolution in cataract surgery begun by Ridley in 1949. As with Ridley's lenses, the first generation of foldable IOLs were poorly manufactured and suffered many decentrations after insertion. Subsequent generations are thinner, have better haptics to stabilise the optic in the eye and have greater biocompatibility.

Most remarkably, phakoemulsification has led to the routinisation of cataract extraction and raised the prospect of cataracts being removed by trained nursing staff. The bottleneck represented by the delicate skill of the surgeon was to a degree replaced by standardised, mechanised and replicable practice; and the economics of the procedure were transformed as the operation became an ambulatory procedure performed in a few hours. Phako also ensured the dominance of the Ridley posterior chamber approach and the return of the ECCE method. By the end of the twentieth century the IOL had become the standard complement to cataract surgery which itself had become one of the most frequently performed outpatient operations in the advanced industrial world (Linebarger et al., 1999). A major survey of the histopathology of IOLs opines that, 'lens implantation is among the safest major procedures in modern surgery' (Apple et al., 1984).

The hip story

Charnley too served in WWII, as an orthopaedic surgeon, and his experiences helped shape his approach when he returned to Manchester and became a consultant under the new NHS. At that time, there were several surgical treatments available for severely arthritic hips. One method, initially favoured by Charnley, was to fix the hip (arthrodesis), limiting motility but reducing the pain caused by movement. More radically, a few surgeons (especially Judet in France and Thompson in the USA) had developed metal replacements for the head of the femur (thigh bone). 'Mould arthroplasties' for insertion between the femur and the socket on the pelvis with which the femur's head articulated (see Klenerman, 2002), were especially associated with Smith Peterson in Boston. But in both kinds of prosthetic operation, the decrease in pain and increase in mobility was limited; and here too, there was a risk of infection and of loosening.

In Britain, just before the war, Phillip Wiles at the Middlesex Hospital London had experimented with replacing both sides of the hip joint (Wiles, 1958). After the war, his disciple G. K. McKee, at a provincial, non-teaching hospital in Norwich, continued with these experiments and in 1951

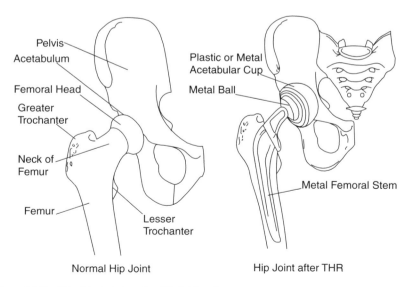

Figure 10.2: The hip joint and artificial implants

implanted the first of a series of THRs with both halves made from metal (Chrome Cobalt). The innovation attracted attention, and a research programme was begun at the Royal National Orthopaedic hospital at Stanmore near London – a collaboration between a surgeon and a graduate engineer (see McKee, 1970; Scales and Wilson, 1969).

Charnley, back in Manchester, was sceptical, but he is said to have gained interest from thinking about the squeaking of a patient with a Judet femoral replacement. As an amateur engineer (like Ridley), he thought the friction could be reduced by using a femoral head that was much smaller than the bone it was replacing. By the mid-1950s, he was looking to develop this idea and was able to establish a unit in a regional orthopaedic TB hospital, at Wrightington, near Wigan, where space had become available as TB had become curable (Swinburn, 1983). This was the sort of opportunity which would not have existed before the NHS provided salaries for consultant surgeons. Charnley extended his skills and range by employing a local fitter and turner, and he also collaborated with a firm of surgical instrument manufacturers who made his prostheses. (He hand-turned his plastic cups in a workshop at his home.) He found cements, expertise and facilities in the Manchester University dental school, and the mechanical engineering department at UMIST – Manchester's technical university (Anderson, 2005). This conjunction of professional tradition, academic research and industrial contacts proved uncommonly productive – eventually.

In 1958 Charnley implanted a THR with a steel head and a plastic acetabular cup. Two kinds of plastic were tried; one proved far too hard and abrasive; the other was Teflon, which appeared more promising and was implanted in about 300 patients, with initial results published in 1961. Charnley thought he had found the materials, designs and technique to make the THR a standard operation – a goal that the IOL did not reach until after phakoemulsification.

But the cups wore, the joints failed, and Charnley was desperate – though the patients stressed the benefits they had received and volunteered for replacements. It could have been the end of Charnley's programme, or at least of the utilisation of plastics rather than metal on metal, but for an accidental contact of a sort which is recurrent in many of the early histories of medical innovations. A salesman for a new industrial plastic, used for gears in local textile machines, called at the hospital and talked with Charnley's technician. The new material proved to wear much less than Teflon, first on the mechanical test rig and then in the body. Charnley resumed his operations, but did not publish again until the late 1960s, when the implants had lasted for more than five years. When this series became publicly known, it was evident that the results were better than other experimental THRs, and the more usual hemiarthroplasties. The metal and plastic combination, with both parts cemented in, became widely recognised as a 'breakthrough' in joint replacement (Waugh, 1990).

From this point, Charnley began to train others in his technique. His surgical instrument suppliers, Thackrays of Leeds, manufactured his specialist tools as well as the implants. They supplied the sets only to surgeons who had trained at Wrightington and satisfied Charnley as to their competence. The surgeons who so visited, including several Americans, formed a club (the Low Friction Society), as in the IOL case. But if Charnley had assumed that these measures would protect his operation from imitation, he had reckoned without the rapid international take-up. One of his friends, the head of AO, a Swiss company of surgeons who had developed the open treatment of fractures, designed a modification of Charnley's system which became popular in the USA (Schlich, 2002). Some American surgeons copied Charnley's device and then modified it. By the mid-1970s, several designs were in use, mostly using materials similar to Charnley's, but which had originated before or alongside Charnley's, mostly in Britain. One sees then the same kind of innovative variation and exploration of a design space that was prompted by Ridley's IOL. In all cases surgeons worked with manufacturing companies, and sometimes with other, academic or specialist, engineers.

The Charnley hip was not patented, and by the 1980s several American companies were developing hips – most of them with a background in older technologies associated with orthopaedics. Most of the American designs used Chrome Cobalt rather than steel for the femur; they tended to avoid cementing (grouting) the femoral component (for which purpose Charnley had developed the use of dental cement); and they were much more expensive.

In Britain, there was a limited convergence onto Charnley's designs. When Charnley first inserted the plastic cups that would prove successful, McKee in Norwich was using a metal or metal design (Chrome Cobalt), and J. T. Scales of the Royal National Orthopaedic Hospital at Stanmore was experimenting with similar materials. Before Charnley's device proved successful, Stanmore's implants were being inserted into patients, and a metal on metal THR was also being developed at Redhill. (Ring, 1969). Both Stanmore and Redhill hospital later adopted Charnley's plastic; and it was used from 1970 at Exeter where Robin Ling worked with the American firm, Howmedica, to develop a variant of the Charnley design for a surgical approach which many found preferable to Charnley's own (Ling, 2002). This proved successful and is now more popular in Britain than the Charnley original or its later variants from Wrightington. The Stanmore hip also continued, but the other two programmes were discontinued.

In the 1980s, designs and new materials continued to proliferate, especially from American companies. Some proved unacceptable, but a variety of configurations survived, allowing surgeons a wide choice of surgical approach, design features and materials – with or without cement. THRs had replaced hemi-arthroplasties for most arthritic conditions, but the THR techniques were far from standardised – partly because of variation among patients as THRs came to be used on younger and more active patients, especially in the

USA, but chiefly because of surgeons' preferences about the operative approach to the hip. Unlike the IOL case, there was, and is, no 'closure' onto one best design, even for a given patient (Neary and Pickstone, forthcoming).

The design space and medical context

One analytical device which we have used to focus our discussion of the two innovations is that of a 'design space' – a conceptual region around the medical problem to be solved. The front of the eye, or the hip joint and approaches to it, were the contexts in which different solutions were developed. Within these design spaces, flows of inventions and innovations, as it were, explored particular trajectories in terms of improved materials, designs and operative procedures. There were no easy distinctions here between invention, innovation and the subsequent diffusion of innovation – rather there was continuous interplay between invention and the experience gained through application.

In such fields, 'theory' provided little guidance, for it was, and is, hard to predict how materials will behave in the living body. Charnley was convinced of the theoretical advantages of a small femoral head, but larger designs in fact worked as well, as even he came to recognise. Animal experimentation was limited, though the 'models' for hips now include ostriches, and big companies now use sophisticated simulators. Even in 2000, when a newly introduced hip system was found to fail more than others, it proved impossible to discover which of the apparently minor changes had caused the problem (Royal College of Surgeons, 2001).

The design spaces were explored by trial and error, guided by mechanical, physiological and anatomical analyses. For the hip especially, the feed-back time was long – often longer than the intervals between the introduction of new variations (see Neary and Pickstone, forthcoming). For the THR, Charnley took ten years to develop a successful system, following twenty years of other British work. And the interactive sequence continues to the present, with the recent introduction of 'minimally invasive' approaches which in some ways correspond to phakoemulsification. Current possibilities also include computer-aided design modelling, the use of guidance systems and the beginning of robotics. For the IOL the development process lasted over forty years from the first implant to the establishment of a standardised procedure on a large scale throughout the advanced nations. Though 'closure' is more closely attained for the IOL, even in this case open possibilities remain, including the possibility of injecting 'intelligent' materials into the capsular bag, from which a lens may self-organise.

We see that knowledges and practices grew in autocatalytic fashion, as one problem led to another. Multiple competing solutions were generated and selected, vicariously or by practical trial and error processes. They can be described as an evolutionary adaptive process, constrained and encouraged by instituted relationships that co-evolved with the growth of knowledge and its application.

In part, this learning was between programmes, as when previous British programmes took up Charnley's use of plastic, or ophthalmic surgeons worldwide adopted phakoemusification. In other cases, these open interactions are less striking than the learning and incremental improvement within a single group. Charnley's own work was remarkable for the systematic approach maintained over decades, covering materials, designs, instruments, cement, infection-reduction – not to mention the training of surgeons and the production of well-illustrated handbooks. Charnley systematically collected data on the fates of his prostheses, including retrieval from patients who had died. We will return below to the question of innovation systems, but Charnley's and Ridley's systematicity will also remind us to be careful about defining the 'objects' of our studies, and to see the wider context of the 'design-space.'

In this wider perspective, we need to remind ourselves that prostheses are not, in principle, the only way of treating cataracts or artificial hips. How much research is there, or should there be, on possible medicines, or on means of prevention? That problem-set is unlikely to be taken on by a devices company, or indeed by operating surgeons, but it should be an issue for the medical professions more generally, and for politics and governments.

Innovation systems

Our two innovations were produced primarily by surgeon inventors, but they did not act alone; they collaborated with specialist companies which were in some sense craft-based. Then more collaborations were added, and more groups entered the field. Companies developed specialist departments and alliances; and a whole raft of agencies, knowledge, skills and resources came into play. This kind of development is sometimes called a distributed innovation system. To medical historians it seems like life as normal (see Pickstone, 1992; Lawrence, 1994; Stanton, 2002; Anderson and Timmermann, forthcoming 2006); but economists see it by contrast with research intensive companies who develop products *in-house*, combining their own skills and resources as necessary.

An emphasis on the distributed nature of invention and innovation processes has already had an impact on the study of medical innovation. Blume's (1995) study of the cochlear implant is concerned with the development of institutional structures to evaluate the feasibility of new devices when their efficacy is strongly contested. The subsequent work of Gelijns and Rosenberg (1999) on CRT scanners, magnetic resonance imaging and endoscopy, makes a clear distinction between the conditions of invention and the conditions that influence the translation of devices into a commercially viable industry. In their account, the strength of local science activity, intellectual property regimes and the characteristics of health care systems play the key explanatory roles. Thus the mix of public and private provision, the patterns of regulation, the conditions surrounding medical experimentation all influence the emergence of particular invention and innovation systems (see Pickstone, 1992).

The two cases reported here speak to the wider question of the systemic nature of innovation in modern economies, and to the crucial interactions that take place between clinicians in hospital contexts and for-profit firms seeking new market opportunities. Though the systems metaphor suggests a degree of self-organisation in the process connecting different organisations and individuals in an innovation system, it also suggests an ongoing element of self-transformation as the flow of innovation tracks an emerging problem sequence. In both our cases, we see the system shifting over decades from contingent connections centred on professionals, to organised programmes centred on companies. Some innovative techniques may be transformed because they are taken over by companies already competent in the relevant field. But in our cases, certainly for hips, the company elements developed along with the innovation, often as special divisions of more general companies. In some cases, these divisions split off, and some were bought by pharmaceutical combines.

But that phenomenon points to a third level, beyond the new device and the new division or company – namely, the development of a new *industry* for medical devices and techniques, alongside and variously linked with pharmaceuticals, but mostly built up after 1950, when pharmaceutical companies were already well established as the major centres of pharmacological innovation.

Surgical innovations

Unlike pharmacological innovation, the development of these medical devices and techniques was professionally based; here, specifically, in surgery. That surgery was professionalised separately from 'physic' or 'internal medicine' is a matter of long history, but it means that modern medicine has large groups of highly trained specialists focused on manual techniques rather than on diseases or pharmaceutical remedies – and in some cases, as in ours, the surgical dexterity and spatial imagination extended to artistic ability and a fondness for crafts (Waugh, 1990). By 1950, in both the US and the UK, eye surgery and orthopaedics were special surgical fields. Both were historically concerned with deformities and their correction, especially in the young, as well as with accidents and specific diseases or disabilities, especially in the old. Both specialisms also had long historical associations with craftsmen and small commercial companies – for the provision of callipers and crutches, etc., and for spectacles of various kinds. Additionally, like most other kinds of surgery, there were associations with companies supplying surgical instruments (and maybe diagnostic devices such as ophthalmoscopes).

Most surgical *operations* were devised and refined by particular surgeons, taught to students and professional visitors, and publicised in professional journals. 'Secret remedies', whether medical or surgical, were scorned by the liberal professions. However, surgeons and/or the small manufacturing companies could patent instruments or devices – which were advertised in catalogues and in professional journals.

Until after WWII, most new operations were tried out on corpses and/or a few animals – and then developed through use on patients. Results were reported, but usually as case-series with little or no statistical analysis. If a technique looked promising, an interested surgeon would often visit the originating clinic to see a few such operations. Then he would try it himself (there were very few women). It was well recognised that surgical skills and techniques were not well conveyed in writing, that direct observation was important, and that even expert surgeons needed to practise operations before they became really competent. From the late nineteenth century, elite surgeons organised 'visiting clubs' so they could attend each other's operations and learn from demonstrations, observation or discussions.

After WWII, it seems, collaboration with companies became more common, if far from routine. And in Britain, where the NHS and the increased state-funding for medical schools meant that surgeons did not have to depend on their private practice, the new salaries increased the possibilities for organised research. Charnley, as a young surgeon in Manchester University, had undertaken systematic research, e.g. on compression arthrodiesis, by which he learned analytical techniques for osteo-histology and for the mechanics of joints (Schlich, 2002).

Many such research collaborations, whether in Universities or with industry, were focused on innovation, but they were often viewed sceptically by fellow professionals, and for good reason. Ridley's IOL and Charnley's THR were radical alternatives to established practice; they placed great demands on the skill of the surgeon and created major risks during and after the operation. Because neither cataract nor hip surgery is a theoretically grounded science, it is entirely reasonable that experience should dominate the world-view of its practitioners and that professional reaction is conservative (see Martin, 2000).

In pharmaceuticals, by about 1960, the randomised clinical trial was being accepted as a template for experimentation; and prompted in Britain by the thalidomide failure, such trials became required during the 1960s. But that was not the case in surgery, where, at least to 1976, the innovation patterns depended largely on surgeons. The techniques and prostheses were taken up by more surgeons in more places; and simultaneously, the innovation was developed for a wider range of patients. At all stages, some risk is involved, but the cost of progress for many may be born especially by the early patients. This is precisely the dilemma that the rules and norms of the profession are meant to deal with and these rules, as accumulated social capital sunk in the profession, will constrain and channel the acceptance of new methods, and make life difficult for innovators. But equally, as we have seen, no account can properly ignore the many patients willing to endure the risks of a new and experimental treatment and so benefit future generations.

Professions, industry and government regulation

We have seen that in both our cases, the initial collaborations of surgeons and companies were driven by surgeons with clear professional goals, who sought advice and assistance about materials and manufacture of devices and equipment. But over thirty years or so, the balance shifted, especially in the USA. In both our cases, and especially from the 1980s, the provision of equipment and the training of practitioners became part of the commercial strategies of the large companies that now dominated the industry.

In the 1960s, the growing use of IOLs in the United States prompted mounting criticism and calls for closer monitoring and regulation of devices. In 1969 implant surgeons in Miami had declared a moratorium on the use of IOLs, to allow them to examine surgical results and decide whether the procedure was safe enough to continue (Jaffe, 1999). They resumed, but this did not mark the end of concerns about IOLs. Broader changes in societal attitudes towards professionals and large corporations were reflected in the emergence of public interest and consumer groups, such as that headed by Ralph Nader. A highly charged debate 'placed ophthalmology in the eye of a surgical storm' and 'polarized the American ophthalmic community like nothing else in recent memory' (both quoted in Jaffe, 1999). It resulted in the extension of the regulatory powers of the FDA to include IOLs in the 1976 Medical Device Amendments. At the same time, the FDA was mandated by the US Congress to institute an immediate study into IOLs whilst not interfering with the availability of the lenses. In December 1981, after the largest clinical study on IOLs ever conducted, the Rayner-designed and manufactured Choyce MkVIII and MkIX lenses became the first IOLs to be approved by the FDA as safe and effective (Rayner, 1999). Paradoxically perhaps, such extensions of the regulatory regime underpinned the growing acceptance of the IOL procedure. Regulation no doubt constrained sharp practice, but it also helped institute the market by adding an implicit minimum quality mark.

For THR also, the British innovations were picked up by American surgeons, who formed alliances with orthopaedic companies to develop particular variants that would feature at major orthopaedic conferences where manufactures displayed equipment; they were also influential through training programmes of various kinds. THRs, too, were included under FDA regulation after 1976, and worries about such regulation were one reason for American suspicion of cemented hips and plastic hip-cups.

Prosthetic materials can be tested in animals for bio-compatibility, and some pioneers including Charnley had inserted plastics under their own skin to guard against immediate reactions. But it is much harder to safeguard patients against possible long-term effects, whether of materials or new designs. Indeed, it is now clear for THRs that minor changes of design and materials can have major effects, which cannot be foreseen. Testing on animals or on simulators, can reduce the risk, but even the most elaborate of recent simulation cannot fully reproduce real-life conditions.

The traditional approach of careful professionals was to introduce changes incrementally and collect data from follow-ups, including devices removed at replacement or at death. But many surgeons were not so careful or systematic, and the proliferation of manufacturers and designs from the 1980s accentuated the problem. In 1995, 62 primary hip replacements manufactured by 19 companies were identified in the UK (Murray et al., 1995).

Regulations in both countries were tightened in the 1990s with the passing of the Safe Medical Devices Act of 1990 in the United States and the establishment of the Medical Devices Agency in the UK. Another approach was to improve the collection of long-term data. A national THR register was introduced in 2003 in Britain, based on regional precedent and a national register in Sweden. From NICE reports and regional registers it is clear that some of the older (and cheaper) models, albeit updated, remain the most reliable and cost effective (Fitzpatrick, 1998; NICE, 2000). But in the *Evening Standard* in 2002, a British surgeon was reported as believing that at least twenty models in the UK could be dangerous (Morris, 2002).

In the USA in the 1990s, as the major companies came to offer designs across the accepted range of THRs, the securing of 'surgeon loyalty' became more dependent on sponsoring young surgeons in various ways, enabling them to train and research using the products of a particular company (Sarmiento, 2003). Exclusive arrangements with hospitals have also become more important, as hospitals sought the best 'deals'. If a surgeon works in a particular hospital s/he may expect to use materials from a particular manufacturer; if the hospital is a major teaching and research institution, then it may tend to become a development site for a particular company. Some leading surgeons have expressed unease about the loss of professional independence. In early 2005, the major hip companies were subpoenaed by the US justice department whose investigations of trading practices in the pharmaceuticals industry have now spread to medical device companies.

In Britain, surgeon-company connections have not been so financially driven, but in May 2004, the Public Accounts Commission reported that 'half of (NHS hospital) Trusts are offered incentives to introduce hip prostheses they would otherwise not purchase and nearly 10 per cent of consultants had accepted incentives from manufacturers mainly in the form of international travel for training purposes' (McCartney, 2004).

Conclusions and national differences

Hips and eyes provide remarkably similar stories of transformational medical innovations on both sides of the Atlantic. Both cases speak to a number of important characteristics of the health care systems in the UK and USA. An economist will note in both cases the apparent paradox that progress in medicine is associated with productivity gains *and* increases in medical budgets. The pre-existing state of knowledge had effectively ruled out treatment on more than an experimental scale, so that the new methods made possible very large productivity gains in terms of quality of life. But as demand was progressively stimulated, extension of treatment made claims on medical budgets and created pressure for their expansion. A state-run health service in the UK managed the process very differently from the private insurance based, decentralised, mixed economy in the USA. The UK remained more economical, but despite its lead in invention and innovation it lost out to the USA in terms of the pace of diffusion and commercialisation.

In the case of the THR, the British innovations, initially driven from within the NHS, involved a limited range of designs, using relatively cheap materials and emphasising feedback of results. There were initial failures, but some of the early designs have proved the most reliable. American developments, which were sooner and more strongly directed by the industrial companies, emphasised expensive materials, novelty and complexity. The profusion of models, especially from the 1980s, meant that assessment was more difficult, and increased competition between a few companies now means that the relations between companies and surgeons is giving rise to concern – in the profession and among regulatory authorities.

But in explaining these trajectories by national contexts, we must also be aware of possible converse effects. The business of prosthetic surgery is now so large and international, especially for joints, that arrangements with companies may be shaping the development of the national medical systems, as when the NHS contracts with companies for special centres to reduce waiting lists for health operations. The effects of particular medical innovations, we may conclude, can extend far beyond their technical 'design spaces' (see Pollock, 2004).

11
Access, Agency and Normality: the Wheelchair and the Internet as Mediators of Disability

Susie Parr, Nick Watson and Brian Woods

Introduction

Since the late 1970s, the recognition that disability is a matter of politics rather than simply a 'problem' of medicine has gained much ground. Similarly, the idea that technology is a political rather than a neutral tool also grew over much the same period (Pinch and Bijker, 1984; Winner, 1980). Yet, despite both engaging with important questions about patterns of power, authority and the human experience, these two fields of study have largely remained separate. In an endeavour to bring them together, this chapter, based on two projects from the IHT Programme, will draw an analogy between wheelchairs and Internet technologies to explore issues of access, control and the autonomy of disabled people. While these are two very different technologies, and the authors bring two very different styles of analysis to bear on the problems, the common thread that ties the two together is that both Internet and wheelchair technologies are sites of political conflict.

Wheelchairs: disconnections and reconnections

Despite being a well-established technology, with a history of at least 300 years (Kamenetz, 1969), wheelchairs remain distant from the status of everyday objects – sitting uneasily between the mundane and the exotic: simultaneously recognisable and yet alien. Though accustomed to their appearance and to a lesser extent their function, most people are unaccustomed with wheelchair use. Disconnected from the everyday experiences of the majority, wheelchairs are machines with which the mass only has fleeting contact. This, along with the inextricable ties between wheelchairs and injury/illness, has had the effect of rendering the technology synonymous with loss, tragedy, passivity and dependency. Wheelchairs are often viewed with trepidation: as machines that disable, confine, and remove from their occupant a state of independence – as medical devices that doctors prescribed only to the sick, the wounded or the elderly.

The status of wheelchairs as medical device probably began in the late eighteenth century, when various machines started to appear in surgical and medical instrument catalogues as vehicles to transport patients. Nonetheless, until the early twentieth century, wheeled chairs occupied a position between a mode of transport for the wealthy and a medical apparatus for injured, sick and/or disabled people. As this intersection dissolved and medical experts took control over illness, disease and disability, wheeled chairs became primarily associated with medicine. The violent consequences of two World Wars also produced an intensification of the enrolment and mobilisation of medical professionals by states across Europe and North America to act as gatekeepers to state provision for disabled veterans, which further advanced a hegemony that classified disability as a medical condition (Stone, 1984) and consequently defined wheelchairs as purely medical devices. As the state became involved in the supply and provision of wheelchairs, they further legitimatised medical control over wheelchairs.

By the end of the twentieth century, various experts made decisions about what type of wheelchair a disabled person should have. Indeed, the association of wheelchairs with medical or health products became so ingrained that the assumption that medical experts rather than wheelchair users should make the buying decision was commonplace (Audit Commission, 2000; Ohras et al., 1997; Phillips and Nicosoa, 1992).

Ironically, though, wheelchairs attracted little kudos within the medical community: seen at best as nothing but dull technologies unworthy of serious attention or at worse, as symbols of failure (Munro, 1949; Abramson, 1950; Hoberman et al., 1952). With its concentration on the cure or alleviation of impairment, medical and rehabilitation ideology and practice constituted and defined what it meant to be 'able-bodied'. The very term, *rehabilitation*, implied 'returning to a *prior* situation' (Stiker, 1999: 122). It was in itself a reference to a norm. This notion of normalcy and its strong association with the medical-industrial construction of disability, developed in earnest following each of the two World Wars. Its initial goal of rehabilitation through technology was to reconstruct and reintegrate maimed veterans into the industrial body, into the productive social machinery: to make disabled people able and useful to both state and society (Bourke, 1996). Hence, the traditional technologies of rehabilitation were the orthoses, the prosthesis, the calliper, the brace, or the crutch – the material forms of the technology based on the notion that what was 'lost' could be replaced or augmented. As Ted Anderson told readers of *Paraplegia News* in 1956, the immediate years following World War II were 'when the popular fallacy of the magnanimous value of crutch-walking seemed to reach its vainglorious peak' (Anderson, 1956: 8). Wheelchair use symbolised either the failure of medicine to find a cure, and/or that the wheelchair user had given up on rehabilitation: an act that countered an ideology, which deemed it the 'duty' of disabled people to adjust themselves to society.

Pat Corby, a medical social worker, interviewed during fieldwork for this project, recollected this normalisation process when she contracted poliomyelitis in 1948:

> When the doctor says something, you are in a very vulnerable position. I mean, I was far more knowledgeable than most patients are and far more used to working with doctors to challenge them, but you know you are very vulnerable. You presume that they know what they are talking about, and when I think about the amount of time and energy, nervous energy, physical energy, that I expended struggling to walk down the ward on two full length callipers, [and] to what purpose? I couldn't lift anything, I couldn't carry anything, I couldn't get over a carpet, let alone anything else, it was the [physiotherapists] and the doctors, they have got a fetish about walking ... but it would have been much [more] realistic to help me cope with being in a wheelchair than anything else. [...] It was ages before I could walk the length of the ward and all my friends were so relieved when I did, because it is not easy to manhandle anyone in callipers you know, they are heavy and they awkward, they are terribly uncomfortable to wear, what the hell was the point, you tell me. (Corby, 2002)

The medical explanation of disability (with its focus on the causal role of impairment) individualised and privatised disability, leading to both marginalisation and social exclusion. It created the conditions under which disability became synonymous with personal tragedy (Oliver, 1990; Barnes, 1991): within such a milieu the claim that wheelchair users could be happy or proud of who they are, challenged and threatened the hegemonic meaning of what it meant to be a disabled person (Swain and French, 2000). Susan Sygall (1998) Cofounder and Director of Berkeley Outreach Recreation Program and Mobility International USA, recalled her experience of these assumptions during the telling of a story about the time she and her friend opened a lemonade stand on Second Avenue and 34th Street in Manhattan. Declaring that it took a great deal of courage for them to do this, ultimately their anxiety turned to anger because people would walk by, see two 18-year old girls in wheelchairs and throw money at them without buying any of the lemonade. The implicit assumption was that because they were in wheelchairs, they were in need of charity.

Segal's anecdote also reveals how assumptions about wheelchairs could often make the user and their intentions invisible. This experience of invisibility appears repeatedly in disabled people's accounts – in particular, of having to endure being ignored while people would talk about or to them through the attendant pushing the wheelchair. Not that invisibility was always a disadvantage. John Hockenberry (2001), for example, tells of how his wheelchair enabled him to be the only journalist that got through a Gaza checkpoint without being stopped despite the area being under curfew.

In terms of wheelchair technology, it is evident that for most of the twentieth century the implicit design assumption was that the user would be housebound or institutionalised. Up until the 1950s, few wheelchairs facilitated independent mobility outdoors: most were simply machines designed to transport a 'patient' from one place to another.

Woolgar (1991) and Akrich (1992) have both observed how engineers often configure or inscribe users and contexts into technological design. Akrich in particular has also shown how assumptions about users are 'inscribed' into technologies so that the finished artefact ends up embodying the values, beliefs, attitudes and norms that potential users are suppose to have. Those designs on offer in medical appliance catalogues during the 1930s and 1940s differed little in configuration from those on offer at the end of the nineteenth century. Wheelchairs were mainly made from wood, most weighed between 70 lbs to 100 lbs and nearly all occupant-propelled wheelchairs had the propelling wheels situated at the front and the castor/s at the rear, the advantages of which were better manoeuvrability in tight spaces and the means to go straight through a doorway without having to first line-up: criteria best suited for indoor use. Once outdoors, front-propelling wheels were useless. The rear castor/s made it impossible to tip and balance these types of wheelchair, which prevented progress up kerbs or up/down steps, and because most were non-folding, assimilation with other forms of transport was near impossible.

Woolgar (1991) also suggested that the 'configuring' process not only defines user identity, but also sets constraints on future action. It is clear that the disconnections of wheelchairs from the mainstream have long shaped the public lives of wheelchair users and cause considerable micro (Cahill & Eggleston, 1994) and macro (Campbell and Oliver, 1996) difficulties. The hegemonic meaning that historically tied wheelchairs to injury and/or illness not only misconstrued the technology as simply a medical device it also had wider consequences that actively disable wheelchair users in a myriad of different ways. It resulted in a lack of understanding on the part of agencies that supply wheelchairs and companies that manufacture them about how they were actually used and it led to the neglect of public transport systems to build-in the wheelchair and the failure of architecture and town planning to account for wheelchair use. Throughout the history of wheelchairs, it is clear that the state and its agents, architects, town planners, and designers and managers of transport systems continually misunderstood the issue of wheelchair access, or actively resisted its implementation (Watson and Woods, 2005).

The low numbers of disabled people in public circulation was a common explanation for the denial of wheelchair access. For example, during discussions on the Chronically Sick and Disabled Persons Bill about access to public transport, the National Bus Company claimed in 1971 that disabled passengers were:

> [A] relatively small part of the totality of bus passengers [and as such, it] would be unrealistic to expect operators to provide special facilities which could cost them money and would be unlikely to increase demand. (Cane, 1971)

Similarly, the American Public Transport Association also fiercely fought Section 504 of the 1973 Rehabilitation Act (part of which required mass transit companies to make new buses accessible) claiming in 1983 that there were too few disabled people to benefit from accessible buses, because the majority of them could not get to bus stops in the first place. Occasionally such arguments bordered on the ridiculous. When confronted with pressure not to discriminate against wheelchair users wanting to fly, the US Airlines Pilots Association and the Airlines Stewardesses Association demanded in 1973 that every flight carrying a wheelchair user 'must be met by rescue, fire and ambulance equipment on the runway' and that a 'Red Cross should be painted' at wheelchair evacuation points (Jacobson, 1973: 2).

Yet, wheelchair history is simultaneously about reconnection: about the attempts of users, disability activists, medical professionals and other important actors to challenge the dominant meanings attached to the technology and reconnect wheelchairs to the mainstream. In part, this involved the politics of access, but reconnection was more complex than adding ramps – it concerned the relations between technology and power. It was about challenges to notions of the norm, about reconfigurations and new readings of the technology, the user, the use environment, and its cultural space – it was about the politicisation of both the user and the machine.

As Rose and Blume (2003) have highlighted, users, contrary to Woolgar's view, can resist their configuration, reject the assumptions made of them by designers, and endeavour to redefine their relations to the technology. The retranslation of wheelchairs from medical device to tools of independence, instruments of politics, and agents of political change began in earnest during the post-war years. Whereas World War I produced colossal numbers of soldiers returning with missing limbs, the technical solution to which was the prosthesis, World War II and the mass production of penicillin by 1945 resulted in the rise of a new constituency of *active* wheelchair users: people surviving with spinal cord injuries. In conjunction, the burgeoning of state organised resettlement and rehabilitation services (including rehabilitation engineering) and a shift in rehabilitation practices and philosophy, and the increased mobilisation of a disabled people into a political movement, raised expectations among this group of being independent (Woods and Watson, 2004b; Hobson, 2002).

From these social conditions emerged tubular-steel, folding wheelchairs, powered wheelchairs and eventually ultra-lightweight rigid-frame and folding wheelchairs. With these important innovations came the possibility of and the increase in demand for independent mobility. Travel and access during the last half of the twentieth century became more than just an aspiration and many wheelchair users were able to more readily leave their homes and partake in mainstream activities for the first time (Woods and Watson, 2004a; Watson and Woods, 2005).

In contrast to the medical construct outlined above, many wheelchair users began to see the technology as a means by which they could gain their independence. Elizabeth Briggs recalled during interview her experience on getting her first powered wheelchair as a child:

> I mean that was when my independence started really, or as my mother said, I was changed from a blob in a wheelchair to a thinking child. ... I became a normal child in a normal society, whereas before I had to rely on my [family] to take me somewhere. I could escape. I threatened to leave home one day. ... I got as far as the first kerb and couldn't get any further, but at least I had the experience. (Briggs, 2001)

Powered wheelchair technology brought with it the possibility of autonomy for many disabled people. Suzanne O'Hara, Director of the Berkeley Disabled Students' Program, 1988–1992, described the time when she was first introduced to a powered wheelchair in the summer of 1971 as:

> [One] of the most revolutionary experiences of my life. ... To go from being pushed ... to being able to control where I was going and the person with me was in a strictly social role, I found that exhilarating. ... It kind-of opened up a whole world of being able to live on my own. (O'Hara, 2002)

Ed Roberts (1994), a founder of both the Berkeley Physically Disabled Students Program and the Center for Independent Living movement, described the impact of his powered wheelchair in a similar vein recollecting how when in a pushchair, he was a nonentity, but once he was out alone in his powered wheelchair he realised that people had to confront him. For Roberts this was a powerful political moment.

Importantly, the ascendancy of these designs was also an effect of wheelchair users' increasingly taking control over the technology. Wheelchair athletes, disability activists, engineers, entrepreneurs all tinkered with and subtly altered their wheelchairs during the post-war years and challenged dominant thinking about wheelchair design. Eventually this tinkering produced some important innovations in both manual and powered wheelchairs in the last quarter of the twentieth century, in particular, more robust and reliable powered wheelchair technology and manual wheelchairs that weighed less than 10 lbs (Woods and Watson, 2003; Watson and Woods, 2005).

Changes in wheelchair design were important in the realisation of independence, but as important was the necessity to reconfigure the user environment. The Invalid Tricycle Association (ITA) in Britain and the Paralyzed Veterans of America (PVA) in the US had an important influence on the politics of wheelchair access in the 1950s. Initially, both groups took on largely similar strategies in the promotion of new building standards to incorporate wheelchair use (Gutman and Gutman, 1968; Goldsmith, 1963). However, both campaigns remained relatively conservative in their approach and outlook. Despite the passing of the Chronically Sick and Disabled Persons Act 1970 in Britain (which included provisions to encourage 'reasonable' adjustments to facilitate access for disabled people) and the Architectural Barriers Act 1968 in the US (which marked one of the first US efforts to ensure access to the built environment) legislation was mostly ineffectual and compliance inconsistent.

Towards the end of the 1960s and into the 1970s, the disabled people's movement in general and the campaign for wheelchair access in particular adopted a more radical edge. Influenced by the civil rights movement and the tactics of the student, free speech and anti-war movements, disabled people came to understand their own social circumstances as an effect of discrimination and an absence of civil rights (Barnartt and Scotch, 2001). Continuing the struggle begun in the 1950s, a variety of groups of disabled people organised around access to the built environment and public transport. Substituting political negotiation with direct action, they employed civil disobedience and confrontation to force societal change. Wheelchairs now became both symbols of and instruments for political action. The potency of wheelchairs as a political symbol was well recognised and had certainly been utilised since the late 1960s if not before. The odd juxtaposition of wheelchairs (objects normally associated with dependency) and protest (disabled people using them to 'march,' occupy buildings, or block buses), transformed a previously invisible object into an extremely visible symbol of disability politics. More importantly however, these early demonstrations, although small in number, were also very powerful in terms of making the participants aware of their own political strength (Grimes, 1998 and 2002; O'Hara, 2002; Fuss, 1997; Billings, 1998), which resulted in the 'wheelchair lobby' becoming relatively powerful. Wheelchairs as physical things, political objects and as cultural symbols thus produced legislative responses that have reshaped our communities and buildings to incorporate their use. Indeed, many actors and organisations translated *disability access* as wheelchair access. This issue of access is, of course, a defining characteristic of the Internet, and it is to this that we now turn our attention, drawing on material from the second research project.

The Internet: technology for all?

The Internet has been conceptualised and represented as a primary means of opening up information and services equally to all. The UK government set the target of ensuring 'universal access to the Internet' by 2005. Government health, social and welfare services are increasingly apparent on the Internet (www.ukonline.gov.uk). The availability of quality information is on the modernisation agenda, and integral to the improvement of health services (Department of Health, 1998b). As well as providing opportunities for telemedicine, information and computer technologies can provide the most effective and efficient ways of disseminating information to the public and to practitioners.

Yet, despite their proliferation and promotion, Internet-based technologies are not yet equally accessible to all groups. For ethnic minority members, older people, young people and disabled people, high potential gains from Internet-based health technologies combine with unequal access and skills. While the Internet brings potential for reducing inequalities (for example ensuring the distribution of quality information, gaining online support and advice), there is evidence that these technologies are not adequately matched to the needs, skills and resources of all potential users (Hughes et al., 2002).

There has been sustained effort to enhance the accessibility of Internet technologies for disabled people. A number of web accessibility guides have been developed. The most widely accepted of these, the Web Content Accessibility Guidelines 1.0 (World Wide Web Consortium, 1999) outlines features of web design that meet the needs of some people with different access needs. However, the focus of this specialised design practice falls on what people with visual, auditory and motor impairments need. For people with communication impairments, who have high potential gains in terms of information, advice and contact, Internet accessibility continues to be inequitable.

One of the IHT Programme projects explored the accessibility of the Internet for people with aphasia, a communication impairment that commonly follows stroke, the single largest cause of disability in the UK adult population. There are at least 300,000 people living with stroke in the UK, of whom 40,000 have aphasia, although some estimates place prevalence much higher. People with aphasia struggle to talk, understand, read and write. Aphasia can vary in type and severity: from occasional difficulty pinpointing a word to intractable inability to use any form of communication. As spoken and written forms of communication are the medium for engagement in personal, social, work, education and leisure pursuits, the impacts of aphasia are profound and far-reaching.

However, aphasia is invisible, unrecognised and poorly understood. People with aphasia often face a dilemma as to whether to keep their impairment hidden or expose it by engaging in a conversation or dialogue. Interestingly, like wheelchair users, people with aphasia commonly have the experience of becoming invisible to others:

> Even the doctor would come in and ask my wife questions, not me. He would come in and ask the wife: 'How is he today?' (in Parr et al., 1997)

The social impacts of aphasia have attracted some research interest, but largely from a clinical perspective. In recent years a social model of aphasia (as opposed to a medical model) has developed (Parr et al., 1997; Simmons Mackie, 2000; Parr 2004), bringing a growing awareness of the disabling barriers faced by people with aphasia and the responsibility of others to alleviate the impact by modifying their own communication and behaviour (Kagan, 1998). The perspectives and emerging identity of people with aphasia have also been explored in line with the social model of disability (Parr et al., 1997; Khosa 2003).

The objectives of this project were to explore the potential of Internet-based health technologies to enable people with aphasia to access information pertaining to health and disability. It also sought to identify the barriers to and facilitators of Internet access for people with communication impairments. Given the isolation and impact on identity and narrative processes brought by communication impairment (Khosa, 2003), the project also investigated the potential of the Internet as a means of self-expression and of creating or joining a virtual community of other people with aphasia, thereby countering isolation and social exclusion. A group of people with aphasia worked on this project, supported by researchers using participative and action research methods. Together, they scrutinised a number of health and disability related websites and documented the barriers and facilitators to Internet access. Drawing on this evidence, they built a prototype website to demonstrate accessible design for this group (www.aphasiahelp.org) and most group members constructed personal web-pages for the site.

The project documented how people with aphasia find the Internet difficult to use. Predictably, this group found their access problems were exacerbated by busy or florid design and layout; multi-tasking demands; font-size, choice of font colour and background; complex navigation; formal, abstract language; complex sentence structure; overwhelming or obscure images and icons. Such access issues are common to people with other impairments.

However, other barriers included more subtle aspects of site design that made the participants feel they were (or more commonly were not) the intended audience. These included the 'tone of voice' of the site; the priorities and balance of the site organisation; the inconsistencies that characterise many sites, even those that profess to be designed for disabled people; site narratives; and participants' own personal purposes and perspectives. The sense of exclusion, when participants were unable to navigate or understand a site or when they felt patronised, was consolidated by feelings of guilt, panic, inadequacy and fatigue. Barriers to Internet access therefore concern issues of design and organisation, but, less straightforwardly, the social interface between the site and the intended audience.

This review of the barriers and facilitators to Internet access is based on an assumption that the desire to seek health information is widely shared. However, the project suggested that this is not necessarily the case. Some participants were indeed interested in finding out information about stroke and aphasia (and other medical and health issues). All described this as a pronounced but often unmet need in the early days following stroke. But others were not interested in accessing information about stroke or exploring interpretations of aphasia, particularly those who operated strong restitution or quest narratives (Frank, 1995). One participant gestured that for him, stroke was in the past. He did not wish to access information about something he had put behind him. The construction of a personal narrative can therefore determine information-seeking priorities. The enquiries people with aphasia bring to the Internet cannot be assumed to be homogeneous.

The feeling of being excluded was exacerbated by the content, ironically perhaps, of charitable sites relating to stroke, and aphasia. These were felt to be *about* people with aphasia, not *for* them. Participants felt excluded by websites demanding a protracted search for information; explanations using medical imagery, terminology and jargon; and suggestions that people contact organisations by phone: a formidable prospect for people who struggle to talk. Such sites precipitated dependence on the support of others.

Obscure language and jargon were even more excluding when they occurred within sites that made some gestures towards increasing accessibility for people with communication impairments. Inconsistency of style and language provoked strong reactions. Following a link on an aphasia website from a homepage written in aphasia-friendly language to a 'research' section, Grace was confronted with a lengthy and densely written academic paper:

> This isn't for somebody with aphasia, it's for dealing with somebody who has aphasia. Not for somebody who can't read long sentences. You lose the meaning of the first bit, words like 'diagnostic', you look it up in the dictionary and you end up with no idea – not good if you're trying to understand your own illness. It makes you feel even more foolish because you're trying to understand and they're not letting you in on it. The rug is pulled from under your feet. You feel you're doing something by logging on, gaining control; this is almost like a big sign saying 'no, you can't come in here'.

Sometimes the very organisations that should understand access failed to accommodate communication impairment, as evidenced in one impenetrable mission statement posted by an organisation supporting disabled people:

> ... provides services for disabled people based on the social model of disability, which aims to identify and remove barriers preventing the full involvement of disabled people in ...

Several disability-related websites were judged to be intended for young people with motor impairments. This impression was underlined by characteristically aspirational narratives. One website showed images of young wheelchair users engaged in various sporting activities, superimposed with the words 'dreams', 'ambition' and 'achievement'. Clashing with participants' pragmatic life-views and accounts of disability, this met with considerable cynicism.

The theme of struggling against adversity was prominent within charitable aphasia and stroke-related websites. Narratives of restitution (Frank, 1995), in which site visitors were encouraged to work hard and persevere in order to recover their lost skills, provoked strong reactions. Some participants were inspired and encouraged, others found the approach simplistic and paternalistic:

> *Stephen:* Erm: [in 'upper class' female voice] 'Oh dear, oh dear, ahhh!' [In own voice] Oh God!
> *Becky:* Kind of a 'there there' attitude, like 'poor you'?
> *Stephen:* (laughs) Yeah. 'Ahhh, ahhh' [gestures patting on head]
> *Becky:* And I get the impression that that doesn't sit too well with you? (laughs)
> *Stephen:* (laughs) Well, erm, yes!

The term 'stroke survivor' (prevalent in some sites) polarised opinion further: some felt it was an appropriate acknowledgement of the seriousness of their experience; others were uneasy with the term but preferred it to 'victim' or 'patient'. Some did not want to be defined by the 'stroke' label at all, regardless of context.

Interestingly, few of the personal stories and accounts of people with aphasia that appeared in stroke and aphasia-related sites showed any evidence of the authors' struggles with communication. They were sanitised and seemed to have been ghost-written. Grammatical and semantic anomalies were edited out and most pages followed standard house style. This is perhaps the equivalent of people who could be independent wheelchair users being given callipers and encouraged to focus on walking. The participants were puzzled by the way in which aphasia 'disappeared' from these sites and commented that there was no clear explanation of how the accounts had been produced. When the time came to produce participants' own web-pages, there was much debate about whether to reveal aphasia, exposing fragmented language and semantic anomalies, or whether to 'tidy up' the contributions. Some opted to show aphasia as it is, others asked the research team to work with them on editing. Everyone agreed that each page should be prefaced with a description of how it was produced. So aphasia was made visible, fore-grounded and addressed in every page.

In summary, project participants with aphasia did not always engage with aphasia, stroke and disability-related websites. Sometimes this was because they did not feel they were the priority audience for the website. Sometimes they felt that the manifestations of aphasia were edited out. When they *did* consider they were being addressed, exclusion occurred when content and style were difficult to process. Finally, and perhaps most significantly, engagement with sites was often contingent on subscribing to a specific perspective on aphasia, stroke and disability. Those who conformed to this orientation responded positively, but others felt they were offered little of value.

Could the Internet provide a virtual community for people isolated by aphasia? The answer to this question seems to be, not surprisingly, yes and no. People with aphasia were transfixed by others' accounts and by the websites that related to their experience ('Fascinating! Fascinating! Wow! Yes!') On several occasions, the group elected to make contact with different organisations and individuals. But there are many barriers to community building. If they were not excluded by websites or alienated by narrative and decided to make contact, people with aphasia often fell at the first hurdle when they had to register, fill in a form, type a password and provide some information about themselves. In the project website, the 'penpals' section attempts to circumvent these hazards.

Conclusions

The above accounts have both explored the complex relations between technology and impairment. Although fundamentally different – wheelchairs are a technology specific to the impaired body, whereas the Internet (at least in its ideal) is a technology for all – there are similarities. In particular, the potential for enhancing disabled people's access (to physical space, information, participation, communication and independence) and ending social exclusion are manifest in the development and use of both technologies. Yet, both studies also show that the realisations of these potentials have only been partial and intermittent. Despite the development of a relatively strong political movement by disabled people, with its associated legislative successes, disability is still individualised, still widely communicated in terms of tragedy and still generally viewed as a medical problem. Indeed, the medicalisation of disability, with its emphasis on the eradication of impairment and/or the normalisation of the impaired body, remains the dominant feature of the sociotechnical landscape. Even today, political action remains a prerequisite for changing the built environment to accommodate mobility impairments and few give thought to the requirements of those with communication impairments when designing the virtual environment. A non-impaired physiology and a narrative of normality are still the primary assumptions of technological design – even in the newest, most rapidly developing technologies.

The studies, however, present two very different stories about the involvement of disabled people in technological production. Whereas wheelchair users became increasingly influential in matters of both wheelchair design and the use environment, people with aphasia have not been a powerful lobbying force over the Internet. In part, a simple explanation lies in the primacy of use. Unlike wheelchairs, disabled people are not the sole, nor even the primary users of the Internet, which means their voice is muted by a myriad of other user demands. Moreover since their voice is often unheard by people without aphasia and since many have difficulty holding onto abstract or complex concepts, this lack of control becomes understandable. Yet, the disparity is more complicated than this explanation may first suggest. While it was the case that some wheelchair users (especially those involved in wheelchair sports) reconfigured the technology themselves, other users (because of severe physical and/or intellectual impairments) had to work with or rely upon non-disabled people to improve wheelchair design.

For those concerned with aphasia, equivalent alliances are emerging. Some clinicians working in rehabilitation are starting to suggest their own reconfiguration, from authority to resource and advocate (Simmons Mackie, 2000; Elman et al., 2003). Within the voluntary sector in the UK, Canada and the USA the practicalities of involvement and partnership with people who have aphasia are being explored and demonstrated (Byng and Hewitt, 2003). Such developments form a cross-wind to prevailing clinical and philanthropic discourses, reframing aphasia in political terms.

The relative successes gained by wheelchair users over the last 50 to 60 years have partially come from the effects of a multitude of actors' attempts to retranslate the technology from a homogeneous thing into a heterogeneous thing. While wheelchairs were arguably a mundane technology, they were (and possibly still are) also alien, in that wheelchair use outside of the medical domain was widely perceived as abnormal. Endeavours to reconnect wheelchairs to the mainstream has had the effects of reshaping the technology both physically and symbolically and producing legislative responses that have restructured our communities, buildings and transport systems to incorporate its use. In changing both the technology and the meanings attached to it, the protagonists made wheelchairs visible to make them acceptable whereby they could once again retreat into the background to become invisible, but normal. Although it continues to be predicated upon struggle, spatial access has been privileged over other forms of access, such as communication access, partly because the latter is an unfamiliar, slippery concept which has yet to be pinned down and operationalised (Parr, 2005). For some users of Internet technologies, the process of gaining visibility is just beginning.

Part 4
Regulation and Evaluation of IHTs

Part 4
Regulation and Evaluation of IHE

12
Understanding the 'Productivity Crisis' in the Pharmaceutical Industry: Over-regulation or Lack of Innovation?

Paul Martin, John Abraham, Courtney Davis and Alison Kraft

Introduction

Concern over productivity and innovation in the modern pharmaceutical industry has a long history (Abraham, 1995: 36–86). However, it first gained prominence in the 1970s and 1980s when the industry itself and various academic/professional commentators on both sides of the Atlantic claimed that government regulatory activity in Western Europe and North America was stifling innovation. One national pharmaceutical industry association after another pointed to how its national regulatory agency was too slow and inefficient in approving new drugs on to the market compared with other countries. For example, during the 1980s, the Association of the British Pharmaceutical Industry (ABPI) complained that the British drug authorities' 'over-regulation' was detrimental to the British economy because drug development work was going abroad. In Germany and Sweden, the pharmaceutical industry also pressed their drug regulatory agencies to accelerate drug approvals and reduce regulatory checks in order to promote innovation (Abraham and Lewis, 2000: 43–79).

Perhaps the most famous articulation of the view that pharmaceutical innovation was a victim of 'over-regulation' was the development of the 'drug lag' thesis by researchers at the industry-funded American Enterprise Institute and Tufts Centre. They attacked the US drug regulatory agency, the Food and Drug Administration (FDA) for depriving American patients and doctors of prescription drug innovations because it took considerable time to conduct checks and was relatively cautious about approving new drugs (Grabowski et al., 1978; Wardell, 1973). The alleged 'drug lag' was between the US, on the one hand, and the UK and some European countries, on the other. The implication of this argument was that these European countries' more 'light-touch' regulation was better than the FDA's approach because it delivered more drug innovations that doctors and patients needed.

Early in the twenty-first century, the spectre of a 'productivity crisis' in the pharmaceutical industry again looms large. In the last ten years or so there has been a decline in the number of marketing authorisations applications for new products in both the EU and the US. For the purposes of measuring new drug productivity and innovation, it is conventional to measure substances whose novelty is defined by technical uniqueness as established by patent acquisition. In this chapter we follow that convention. These substances are known collectively as 'new active substances' (NASs) in Europe and as 'new molecular entities' (NMEs) or new 'biologicals' in the US.

Figure 12.1 shows the annual number of marketing authorisation applications submitted either to the EU's drug regulatory agency, the European Agency for the Evalution of Medicines (EMEA), under its centralised procedure or its mutual recognition procedure for NASs (both new biologicals and NMEs) that were not subject to a previous application. The data, which were compiled by Charles River Associates on behalf of the European Commission, show a marked drop in the number of marketing authorisation applications from 74 in 2001 to 47 in 2003 (Charles River Associates, 2004).

Data compiled by the FDA show a similar downward trend in the number of licensing applications for new drug and biological entities submitted to that agency, with the number of annual licensing applications for NMEs falling from around 45 in 1997 to around 25 in 2003 and the number of new biological licensing applications falling from around 45 in 1995 to under

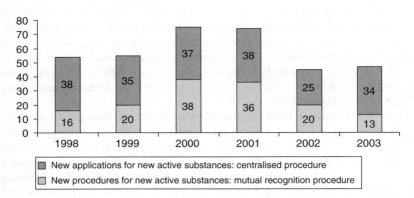

Figure 12.1: Licensing applications for new active substances in the EU under the centralised procedure and the mutual recognition procedure – 1998 to 2003
Source: Charles River Associates, 2004: 16.

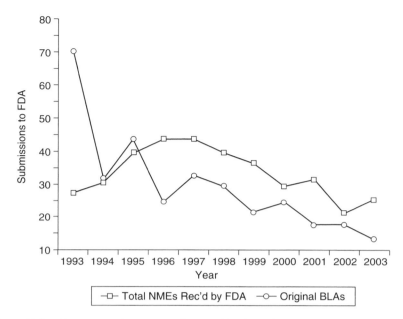

Figure 12.2: 10-year trends in major drug and biological product submissions to the FDA

Source: FDA. Available at http://www.fda.gov/oc/initiatives/criticalpath/nwoodcock0602/woodcock0602.html

5 in 2003. Figure 12.2 illustrates this trend over the ten-year period from 1993 to 2003.

The falling number of applications has been mirrored by a decreasing number of new product approvals and launches in the US and the EU (Charles River Associates, 2004; Turner, 2004). In fact, data show that global NME output has declined (see Figure 12.3 below), leading many commentators to ask whether the industry is now facing a *worldwide* crisis in innovation (CMR International, 2002; Charles River Associates, 2004; FDA, 2004a; CMR International, 2005). Just 26 new molecular entities were launched on the world market in 2003 – fewer than at any time in the last 20 years (CMR International, 2005).

In this chapter, we suggest that the relationship between regulation and innovation is much more complex than the dominant view that 'over-regulation' inhibits innovation. Furthermore, we argue that it is highly implausible that the current 'productivity crisis' in the pharmaceutical industry is due to over-regulation. Rather, we suggest that the industry has become locked into an innovative paradigm focused on incremental

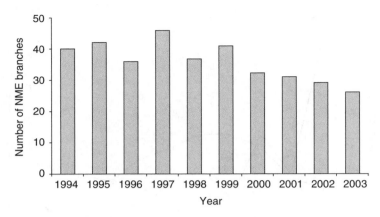

Figure 12.3: Number of NMEs first launched onto the world market (1994–2003)
Source: CMR International, 2005.

innovation and the pursuit of blockbuster markets. This has resulted in the exhaustion of established sources of novelty at the same time as there has been an increase in the complexity of the diseases being targeted. Taken together, this has trapped companies into a pattern of diminishing returns.

The role and nature of regulation

In market economies there exists 'a perpetual tension between the freedoms conferred by the private ownership of productive property and the need to impose communal limits on the exercise of those freedoms' (Hancher and Moran, 1989: 1). Regulation is a central way in which governments attempt to manage that tension. The nature of regulation in market economies is highly contested, as indeed is the nature of the state of which government regulatory authorities are part. Few would question that, for analytical purposes, government regulation may be regarded as state intervention in the market, though, in reality, regulation also creates and circumscribes markets. As such, the 'regulatory state' affects the interests of producers and consumers in the marketplace (Majone, 1994). In particular, regulatory activity impacts on private industry. There is also agreement that regulation is generally introduced because of the inadequacies of an unregulated market.

In the US, government regulation of technological innovation of private industry as a response to market failure dates back to the late nineteenth century with the establishment of a federal agency to regulate the railways. Regulation was seen as a way of promoting efficiency gains. Throughout the twentieth century many other federal agencies responsible for the regulation of private industries were to follow, including the FDA in 1927. By contrast, in Europe, until the 1980s, the predominant response to market failures was to nationalise private industries (Majone 1996: 9–19). However, since the 1980s, many countries in Europe have, in some ways, moved towards an American style of regulation. Moreover, despite claims that the market fails to deliver medicines that patients need, few Western European countries ever nationalised their *pharmaceutical* industries.

According to Freeman (1982: 56–105), industrial innovation is a coupling of technical and market opportunities. As regulation partly defines markets by permitting the sale or exchange of some technologies but not others, there is not necessarily a negative correlation between regulation and innovation. Even if regulation is defined as 'state control over activities regarded as desirable by society' (Majone, 1990: 2), regulatory developments may limit, stimulate or alter the direction of innovation. Moreover, the complex relationship between regulation and innovation may not be unidirectional: scientific and technological innovations can influence regulatory developments (Pieters, 2003). In this context, one should be wary of responses to stalling innovations, which automatically seek to lay the blame on 'over-regulation'.

At its simplest, technological innovation involves discovery *and* the development of that discovery so as to bring it to a market (Jevons, 1992). In the pharmaceutical sector, the dominant definition of a drug discovery is the identification of a NME or NAS, which has therapeutically interesting biological activity (Vos, 1991). Yet even at this early stage of innovation, regulation plays a fundamental role because the status of NME (or NAS) is conferred by patent regulations which define technical novelty.

Regulation plays a much more intricate role in the development phase of drug innovation. The public interest rationale for drug regulation, developed across the western industrialised world by the 1970s, is that it should protect patients from exposure to unsafe or ineffective pharmaceutical products (Abraham, 2002a). Thus, in deciding whether or not patients should be exposed to a newly developed drug it is necessary to make some risk–benefit assessment of the prospective medicine in question. To do this regulators and others turn to the scientific testing of drugs which comprises toxicology, clinical trials and (pseudo-)epidemiology. Regulatory agencies typically require the following battery of tests to be conducted in the development stage of new drugs, though in some circumstances application may be made earlier than completion of phase III trials:

- Preclinical and Non-clinical – necessary animal and bench testing before administration to humans plus start of tests which run concurrently with exposure to humans;
- Phase I trials – first studies of new compound in humans; usually healthy volunteers;
- Phase II trials (Proof of Concept) – first studies in patients to explore efficacy and safety in people who have the target illness/condition;

- Phase III trials – studies in a large patient population to generate more extensive and precise safety and efficacy data for licence application;
- Licensing/ marketing authorisation application;
- Phase IV – post-marketing adverse drug reaction (ADR) reports and studies, if application approved.

While drug regulatory agencies require these tests to be completed, they do not conduct them. Western governments do not conduct any of their own independent tests of new drugs developed by the pharmaceutical industry. The industry is responsible for all drug testing and when the appropriate phases are complete, the company submits its results to the regulatory agency for review and possible marketing approval.

Toxicology is an experimental science conducted mainly in industry. Toxicologists draw on data derived from *in vitro* studies of cell behaviour carried out in glass dishes and on *in vivo* research conducted in whole live animals. Using these two types of data they try to predict the toxicity of a compound in humans. Assuming the compound is not abandoned due to adverse toxicity (or for commercial reasons), the manufacturer then proceeds to test the drug in humans. It may be noted that some, but not all, toxicological tests on a new drug are pre-clinical, that is, completed before clinical testing begins. Some are 'non-clinical' and proceed in parallel with human testing.

The pharmaceutical manufacturers take the lead in organising and conducting human/clinical testing, but unlike toxicological testing, the industry generally involves the wider medical profession at this stage, especially high status medical professionals in academia or teaching hospitals who might be sympathetic to the manufacturers' endeavours. Generally clinical testing takes the form of double-blind controlled clinical trials in which one group of patients receives the new drug and another group receives either a placebo or another group of the same therapeutic class. During the trials neither the patients nor the clinical investigators know who is receiving which 'treatment' until all the results concerning the patients' responses are recorded. Hence, such trials provide comparative data about the toxicity and effectiveness of new drugs in patient groups. However, no regulatory agencies in the EU or North America require pharmaceutical manufacturers to demonstrate that their drug is more efficacious than products already on the market – indeed the new drug may be less so, so long as it demonstrates an 'adequate' level of clinical efficacy (Abraham, 2003).

Both toxicological tests and controlled clinical trials may continue after a medicine has been put on the market. However, at that stage a type of (pseudo-) epidemiological data also becomes available, namely spontaneous reports of ADRs experienced by patients taking the drug either in general practice or during hospital treatment. These data, derived from post-marketing surveillance systems (also known as pharmacovigilance) are collected initially by pharmaceutical companies and doctors who may then report them to regulatory authorities. Pharmaceutical companies are required by law to report ADRs to various degrees in the US and EU, but the arrangements by which doctors make such reports are voluntary (Abraham, 2004). In addition, companies may be requested and/or required to conduct post-marketing trials with a new drug in response to regulators' concern about some safety issue.

All of these regulatory activities offer opportunities for companies to produces drug innovations that are safer and/or more effective than existing therapies.

Impact of regulatory change on drug approvals

Since the 1980s governments have accepted the pharmaceutical industry's demands for regulatory reforms ostensibly aimed at increasing innovation. European governments decided to restructure their drug regulatory authorities in line with the industry's demands. In the late 1980s, the industry proposed that it would be willing to pay the costs of funding drug approvals if this were to result in a 'more efficient service', and called for greater informal consultation between companies and regulators (Anon, 1987; 1988a, b). For example, in 1989, the UK Government accepted the industry's proposal that, in order to make medicines regulation more 'efficient', the entire cost of running the regulatory authority should become dependent on fees paid by pharmaceutical companies for licensing. Many other EU regulatory agencies became similarly restructured (Abraham, 2002b).

Furthermore, this created a situation in which national regulatory agencies compete with each other for fees from pharmaceutical companies in order to maintain revenue into the regulatory institution. As regulators know that speedy approval of new products is one of the industry's objectives, then the competition between European regulatory agencies is significantly driven by a tendency to accelerate the process of regulatory checks on new drugs. In addition, to accommodate the industry's desire for more rapid approval times, strict timescales have been prescribed by the European Commission for regulatory review within EU procedures. The competition for fees from industry means that regulatory agencies have not even been satisfied to meet the European Commission's requirement of approval (or non-approval) within 210 days. For example, the average net in-house assessment times by the UK drug regulatory agency for new drugs fell from 154 working days in 1989 to just 44 days by 1998. The drug regulatory review times for Germany, Sweden, other EU countries and the US also fell dramatically in this period (Abraham and Lewis, 2000: 20; Kessler et al., 1996).

Drug review times accelerated in the US because, by the late 1980s, the FDA was acting on the criticisms from the 'drug lag' thesis and complaints about over-regulation from the industry. In 1992, the Prescription Drug

Users Fee Act (PDUFA), strengthened this perspective with legislation. The PDUFA required that the funding of the FDA change from being entirely independent of the industry to being 15 per cent dependent in fees from companies in exchange for acceleration of the drug review process. Yet, even in the 1970s and 1980s, the claims that the industry's productivity and innovation problems lay in 'over-regulation' were highly implausible. The first hint of this lies in the fact that each national pharmaceutical industry association argued that its regulatory agency hampered innovation in that country. For example, advocates of the 'drug lag' thesis in the US held up the UK regulatory system as a good standard for promoting pharmaceutical innovation at the same time as the ABPI was lambasting it for inhibiting innovation. Of course, all of these assertions about comparative 'over-regulation' could not be true.

In fact, research into the industry's claims at that time would have shown that in the period 1972 to 1983, on average, NMEs came to the UK and/or German markets faster than in France, Italy, Sweden or the US, and that from 1961 to 1985, more new drugs were first marketed in the UK and/or Germany than in Austria, the Benelux countries, the Eastern Bloc, Italy, Scandinavia, Spain, Switzerland or the US (Andersson, 1992). Indeed, in 1988, the UK was found to have the fastest approval times for new drugs in the EU (Anon, 1988c). In this light, the suggestion by the ABPI that drug development might go abroad because of the slowness of the British drug regulatory authorities in reviewing new drugs, and its alleged concomitant damage to innovation, must be viewed with incredulity.

By comparison, during the 1970s and 1980s, the FDA was slower in approving NMEs overall than regulatory authorities in the UK and many other European countries. However, it is now known that a major consequence of this regulatory approach was to prevent many more unsafe new drugs from entering the US market – drugs which were approved in the UK and subsequently had to be withdrawn following injury to patients (Abraham and Davis, 2005). Furthermore, Schweitzer et al. (1996) analysed the approval dates of 34 pharmaceuticals, marketed in the G-7 countries plus Switzerland between 1970 and 1988 and designated especially therapeutically significant (at the time of their approval) by panels of doctors and pharmacists in the US and France. The FDA was found to approve more of these drugs before the UK regulatory authorities and ranked third out of the eight countries in approving these drugs on to the market. This suggests that, while the 'drug lag' thesis may have been correct regarding NMEs taken as a whole during the 1970s and 1980s, it may be largely irrelevant to a discussion about those pharmaceutical innovations that are of most value to patients, health professionals and public health.

If 'over-regulation' was an improbable cause of productivity problems in the pharmaceutical industry in the 1970s and 1980s, it is even less likely to be responsible for the current 'productivity crisis'. Over the last ten years regulators on both sides of the Atlantic have continued with regulatory reforms based on the ideology that over-regulation has been inhibiting innovation. In the US, the Senate Committee report accompanying the Food and Drug Administration Modernisation Act of 1997 asserted that:

> Increases in the time, complexity and cost of bringing new products to market are a growing disincentive to continued investment in the development of innovative new products and a growing incentive for American companies to move research, development, and production abroad, threatening our Nation's continued world leadership in new product development. (US Senate, 1997: 7)

Conversely, within this ideology, shortening drug development and review times and increasing predictability in the drug development process fosters industrial innovation and hence points the way to further reform (Milne, 2000). The Charles River Associates report, commissioned by Enterprise DG of the European Commission, makes explicit that one of the main purposes of accelerated drug reviews in the EU is to encourage innovation (Charles River Associates, 2004: viii).

Over the last ten years, these industry-friendly reforms have certainly been delivered in the EU, US and beyond for both development and review times. Kaitin and Di Masi (2000) found that the average clinical development time for new drugs given priority by the FDA was 48 per cent lower in 1998–99 than in 1990–91. They concluded that one explanation for the recent trend towards decreased clinical development times – as compared to steady increases in development times over the previous three decades – was that the regulatory initiatives introduced in the US had been effective. Also between 1998 and 2003 the FDA's fast-track programme for some new drugs had the intended effect of shortening drug development times – with a reported cut of nearly three years from the average time required to develop a new drug and gain approval (Tufts Center for the Study of Drug Development, 2004: 3).

FDA review times for priority NMEs decreased steadily from 1992. In 1993 the median FDA review time for such drugs was 13.9 months. By 2004 this was just six months. Similarly, median FDA review times for standard NMEs fell from 27.2 months in 1992 to 16 months in 2004 (FDA, 2004b). Performance indicators of the EMEA show that every year from 1995 to 2003 timelines were adhered to, with the average time for the assessment phase of the procedure in 2003 recorded as 190 days compared to a target time of 210 days (EMEA, 2004: 16).

In addition, since 1990 there has been international standardisation of drug testing in order to streamline the drug development process for transnational pharmaceutical companies. Central to this trend has been the International Conference on Harmonisation of Technical Requirements for

Registration of Pharmaceuticals for Human Use (ICH). The key participants are the three pharmaceutical industry associations and three government drug regulatory agencies of the EU, Japan and the US. However, some of its standards have been adopted outside these three territories by countries such as Canada and Australia, and by the World Health Organisation. At the behest of streamlining the industry's innovation process, this international harmonisation has lowered or loosened many safety standards in toxicology, clinical trials and post-marketing surveillance of ADRs (Abraham and Reed, 2001; 2002; 2003).

Nevertheless, despite all these measures over the last twenty years to reduce, streamline and compromise government's regulatory oversight of the industry in order to foster greater pharmaceutical innovation, the industry currently seems to be facing its greatest 'productivity crisis' in modern history. Evidently, the ideology of 'over-regulation', which has held sway for the previous two decades is misguided and one must look for more complex reasons for this 'crisis'.

The changing nature of the drug innovation process: impact on productivity

In order to understand the nature and extent of the industry's productivity crisis, it is important to place the current debate in the context of changes in drug innovation and technology since the 1950s. Following the development of the modern pharmaceutical industry in the decades after WWII, there have been a number of important shifts in both the drug innovation cycle (DIC) and the types of products being developed by the industry. The commercialisation of penicillin marked a watershed in the industry's development, and wartime demand for antibiotics prompted the drug industry's transition to an R&D intensive business (Henderson, Orsenigo and Pisano, 1999). In the 1950s and 60s heavy investment in R&D was increasingly seen as the key strategy for both product innovation and competitive advantage. This provided the rationale for the massive expansion of in-house research that was unbroken until the late 1990s, and which has been perhaps the defining feature of the sector.

In the post-war decades, drug discovery was based on synthetic chemistry and was led by the search for drugs related to antibiotics. By 1962 antibiotics accounted for 25 per cent of total world pharmaceutical sales and the quest for new antibiotics was closely associated with the development of large screening programmes within the industry. In particular, the random screening of libraries of chemical molecules (so called 'molecular roulette') became the dominant method of drug discovery.

Although the emphasis was on so-called random screening, the starting point for the creation of chemical libraries was typically a compound with a known physiological effect. Synthetic chemistry would then be used to create related, modified or derivative compounds. This strategy was attractive, as it increased the likelihood of success and reduced the cost of discovery, but resulted in the creation of a series of so called 'me too' (or congener), products which were often closely related to a previously successful compound. Typically, 'first-in-class' drugs triggered a search amongst competitors for congener products, used for similar therapeutic indications, but which offered marginal improvement on the compounds they imitated.

The 1950s and 1960s was a time of high productivity during which the random screening discovery paradigm gave rise to a number of important new drug classes, including the steroids, the contraceptive pill, antipsychotics, the benzodiazepines, the calcium antagonists and the Beta-blockers This strategy proved to be technically successful and commercially attractive, and in the post-war decades the industry enjoyed high profits and above average returns on investment. At the same time, however, this strategy institutionalised a highly conservative approach to innovation based on incremental improvements on existing products (Drews, 2000), and by the early 1970s over 80 per cent of newly patented pharmaceuticals in the UK were congener drugs (Goodman, 2000). The trend for developing new drugs through modification of successful related products gave rise to a number of drug 'families', exemplified in the sulfa drugs where a single chemical motif within sulphanilamide was the basis for antibiotics, a range of hypoglycaemic agents, diuretics and anti-hypertensive drugs. This convergence on a limited number of chemical structures has remained a central feature of the industry. For example, the 483 drug targets addressed by all approved pharmaceuticals in 1996 could be divided into just six classes, with receptors and enzymes being the main groups (Drews, 2000). Furthermore, the receptor group is dominated by one sub-class (GPCRs), which form the target of 30 per cent of all marketed drugs and constitutes 20 per cent of the top 50 best-selling drugs.

However, during the 1960s the environment in which the industry operated started to change, with much greater emphasis on drug safety, scientific proof of efficacy and state regulation in the wake of the thalidomide disaster. These changes were initiated in the US largely as a result of the Kefauver-Harris Amendments to the existing Food, Drug and Cosmetic Act in 1962 (Vos, 1991; Peltzman, 1973). By the 1970s, the growing statutory regulation of drug development started to seriously impact on the duration and cost of the innovation process, albeit in ways that still remain poorly understood. In particular, pre-clinical testing was expanded (especially toxicity testing) and clinical trials were extended to include larger patient cohorts. Concern became focused on the high attrition rate for candidates in late stage (Phases II and III) clinical trials, in which tightened regulation appeared to be implicated, as product failures at this stage were very costly. As a consequence, renewed emphasis was placed on the discovery and pre-clinical phases in order to improve the selection of drug candidates that

Table 12.1: Productivity in the pharmaceutical industry 1951–1970

Period	Four-year averages of no. of NCEs introduced in US
1951–1954	39.0
1955–1958	42.0
1959–1962	43.5
1963–1966	17.0
1967–1970	15.3
1971–1975	15.0

Source: Bognor, 1996: 95.

would be successfully taken into clinical development. One outcome of these various pressures and changes was a lengthening of the drug innovation cycle, which in turn, increased some of the commercial risks involved.

The rising cost of R&D and a steady decline in the number of NCEs reaching the market (see Table 12.1) drew attention to the problem of diminishing returns from incremental innovation and cast doubt over the long term viability of the random screening discovery paradigm. Amid this climate, the industry started to look for novel discovery techniques. In 1972, new requirements imposed by the FDA brought about a more stringent regulatory regime, which required detailed information about the mechanism of action of all new drug candidates (Bognor, 1996; Abraham, 1995). This coincided with a shift in which greater emphasis came to be placed on the biological target of the drug; referred to as Rational Drug Design (RDD). In particular, it was hoped that RDD would bring about more rigorous selection of candidates for downstream development, reducing attrition in the clinical phases and improving the overall efficiency of the innovation process.

RDD combined biochemical targeting with insights from structural biology and computer modelling, and soon became the dominant approach within pharmaceutical R&D. The adoption of RDD marked an important change in the innovation process, with far greater attention paid to the pathological pathways against which a drug was targeted. New classes of drugs were made possible by this improved understanding of the physiology of pathology, exemplified by a new generation of cardiovascular disease (CVD) therapeutics (Vos, 1991). The first product resulting from RDD was the anti-hypertensive drug Captopril (capoten) launched in 1977 by Bristol Myers Squibb.

The move to RDD in the 1980s coincided with efforts to control health care expenditure and increasing downward pressure on both drug prices and company profits (Bognor, 1996). At the same time, this new approach to drug discovery was costly to undertake and was associated with increased R&D expenditures, raising the minimum effective size thresholds for firms. In 1970 pharmaceutical companies invested, on average, some 12 per cent of total sales in R&D, but by 1994 this had risen to 20 per cent. Significantly, however, the vast increase in R&D budgets that took place during the 1980s and 90s (see Figure 12.4) was not accompanied by an increase in productivity as measured by the number of new NCEs reaching the market (Booth and Zemmel, 2004).

If RDD changed the way in which drug discovery was carried out, it came to prominence at a time when the focus of the pharmaceutical industry was moving away from infectious disease and towards chronic conditions, notably CVD, which were becoming increasingly prevalent. This dovetailed with commercial considerations, with market size becoming increasingly important as firms sought to sustain higher (R&D) costs. This pushed the focus of research in particular directions, towards complex conditions such as CVD. In the early 1960s, anti-infectives and CNS therapies accounted for just under half world drug sales, but by 1993 CVD therapies were the most important class of drugs accounting for 22 per cent of total sales (Goodman, 2000). This and other chronic diseases, such as autoimmune disorders and diabetes, were increasingly seen as providing the most reliable means of generating massive sales and maximum profit. Initially, this strategic shift paid off commercially and was associated with the birth of the 'blockbuster culture', which has come to dominate the industry in the last quarter

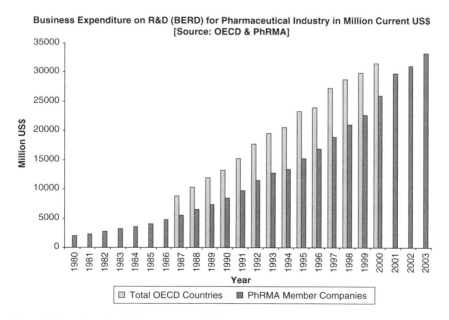

Figure 12.4: Growth of pharmaceutical R&D (1980–2003)
Source: Data from Nightingale and Martin (2004).

century. Between 1993 and 2003 more than thirty newly launched drugs achieved blockbuster status, with sales of more than $1 billion. If the blockbuster model made strategic sense and delivered commercial success, however, it also further sensitised the industry to the dangers of late-stage failure.

The growing importance of blockbuster products has played a key role in the increasing reliance by the first tier of pharmaceutical companies on a few leading drugs – a reliance augmented by investor expectation of blockbuster drugs. However, although the blockbuster strategy may have generated massive revenues, it has done little to alleviate the productivity crisis, nor has it diminished the entrenched preference for congener drugs, which continue to form the basis for many blockbusters. Thus, the industry-wide preference for blockbuster drugs has amplified the trend towards convergent and incremental innovation. In effect, revenues from blockbuster sales belied an underlying situation in which overall productivity – as measured by the number of NCE products/year – has continued to fall. Over time, this has become increasingly apparent (Reichart, 2003) and by the 1990s the innovation problem was being openly referred to in terms of a 'productivity crisis'.

The search for new sources of innovation

In addition to the increased pressure from companies to reduce the regulatory burden and speed up drug approval times, the industry has continued to search for new ways of discovering and, to a lesser extent, developing new products. During the 1980s great interest was expressed in the potential of molecular biology and biotechnology to create new classes of 'biological' products, such as therapeutic proteins, monoclonal antibodies and vaccines. However, in practice the majority of the mainstream pharmaceutical industry adopted a 'wait and see' approach, preferring instead to invest in the shift to RDD (Galambos and Sturchio, 1998). Whilst many molecular biology techniques were rapidly incorporated into established innovation processes, there was relatively little direct investment in new types of biological products, except in cases such as Lilly, where established markets were under threat. Instead, the development of the first wave of biotechnology products, therapeutic proteins, was left to first mover biotechnology companies established in the late 1970s and early 80s (e.g. Genentech, Amgen).

This situation started to change in the mid-1990s, with large firms increasingly seeing the maturing biotechnology sector as an important source of new products. Partly this reflected changes in the biotechnology industry itself. During the late 1980s and early 90s the US biotechnology sector grew rapidly, so that over 1,000 firms were in existence by the middle of the decade, and most of which were focused on human health care. A similar period of rapid growth was experienced in Europe in the late 1990s. At the same time, the technical focus of the biotechnology industry diversified from the first generation of successful biotech products, such as recombinant insulin, EPO and tPA. Increasing attention started to be paid to the application of new biological knowledge to the discovery of the classic small molecule drugs that had been the mainstay of the post-war pharmaceutical industry.

Large incumbent firms responded to the new opportunities presented by the maturing biotechnology sector and pressure to fill their dwindling product pipelines by establishing large number of strategic alliances with biotech SMEs. By the end of the 1990s a new form of heavily networked industrial structure was emerging, with large companies committing as much as 30 per cent of their R&D budget to technology and product development collaborations with smaller companies. Recent data suggests that during the period 1998–2003 the biotechnology sector was the source of as much as 28 per cent of new molecular entities and 85 per cent of new biological entities submitted to the FDA, and 36 per cent of new drug approvals (Kneller, 2005).

The changing structure of the industry also coincided with the sequencing of the human genome and the rise of a genomics sub-sector within the biotechnology industry. Genomics promised to provide a massive increase in the number of new drug targets, many of which appeared to be completely novel, and was developed in large firms by both the creation of in-house research groups, as well as external partnerships. The explosion in the number of potential drug targets could be fully exploited by two other important technological developments that had occurred during the 1990s; combinatorial chemistry and high throughput screening (HTS). Combinatorial chemistry allowed the creation of massive libraries of tens of thousands of chemical variants around a particular molecular structure, and HTS enabled these to be screened against a particular drug target at high speed. This strategy rapidly reduced the cost of generating compound libraries and gave new impetus to large-scale screening. However, whilst the philosophy of RDD, with its emphasis on characterising the drug target was still very influential, the vast majority of the new targets generated by genomics were often poorly characterised, with companies having little understanding of their role in disease pathology. This led to a paradoxical situation where the numbers of targets being investigated by companies has risen from an average of 50 in 1996 to 200 in 2003, but the numbers of scientific papers associated with each target fell from 100 to 8 in the same period (TUFTS, 2006). In other words, genomics has vastly increased the potential for innovation within the industry. At the same time, it has shifted the key innovation bottleneck from the creation of novel small molecule drugs against known targets (chemistry) to the functional characterisation of large numbers of unknown drug targets (biology) against which it is relatively easy to create drug candidates. This can only be achieved through heavy investment in disease biology and is a process that is likely to take many years.

The FDA has also identified another problem that it claims the industry will need to address in order to overcome the productivity crisis. In a White Paper title 'Innovation or Stagnation: Challenge and Opportunity on the Critical Path to New Medical Products' (FDA, 2004a), the Agency argues that new technologies of clinical and preclinical assessment need to be developed in order to tackle the current shortcomings on the drug development process. This is a significant development, with the world's leading regulatory agency arguing that it has been technological failure that has been responsible for the productivity crisis and that the locus of the problem is the development, rather than the discovery phase of the DIC.

Conclusions: understanding the innovation crisis

At the start of this chapter we outlined the influential argument that 'over-regulation' was the main cause of the industry's current productivity crisis. However, the available evidence calls this seriously into question for a number of reasons. Firstly, the decline in productivity started in the 1960s, well before the creation of the current regulatory regime. Secondly, there have been a series of very successful initiatives that have fostered faster regulatory review and market authorisation processes over the last decade. However, these changes took place at a time when the productivity crisis was worsening. It is therefore very questionable if the pharmaceutical industry's productivity problems have ever been due to 'over-regulation'. Regarding the current 'productivity crisis', the evidence suggests that 'over-regulation' is not a factor.

An alternative and more convincing explanation would instead centre on the decreasing returns that arose from the industry's 'lock in' around a paradigm based on incremental innovation based on a limited number of established drug targets. In the 1960s and 70s the commercial benefits of pursuing this 'me too' strategy were attractive, and the turn to RDD in the 1970s and 80s provided it with new life. However, by the late 1980s there were clear signs that it was becoming increasingly costly and time-consuming to create the novel drugs required to treat the more complex diseases that were the basis of blockbuster markets. This combination of the exhaustion of the innovative potential of existing drugs and drug targets, combined with the much more complex task of treating chronic conditions, has created an innovation trap. Despite ever increasing R&D budgets, it has proved very difficult to get out of this trap through the adoption of new and more radical approaches to drug discovery and development.

The innovative potential of genomics and the outsourcing of R&D to small biotechnology and genomics companies may in the long run provide the industry with both a rich source of novel products and the organisational flexibility required to embrace new approaches to innovation. However, this will not happen quickly, as it will require the accumulation of a new body of knowledge about the role of these new drug targets in disease pathways in order for them to be fully exploited.

Furthermore, from a historical perspective, the possibilities for wholesale change within the pharmaceutical industry have, since World War II, grown increasingly difficult for a number of structural reasons. Amongst the factors militating against rapid or through-going change are the acute commercial risks raised by radical change, the dominance of the blockbuster culture, the constraints inherent in the lengthy duration of the DIC, and the way in which the regulatory system has reinforced certain types of innovative strategy. From this perspective, an interesting line of investigation might be to consider that innovation in the industry may have suffered because it has not been sufficiently stimulated by challenging regulatory demands, such as a requirement for new products to demonstrate therapeutic advance. In other words, the productivity crisis is not the result of 'over-regulation', but paradoxically may have been caused, in part, by the failure of regulatory authorities to place sufficient demands on industry.

13
Regulating Hybridity: Policing Pollution in Tissue Engineering and Transpecies Transplantation

Nik Brown, Alex Faulkner, Julie Kent and Mike Michael

Introduction

Conventional boundaries of the material and social worlds are increasingly challenged in late modern society. Human and animal matters have been globally mobilised in a worldwide traffic of scientific, medical and commercial transactions. Whilst human implant technologies have a long history, contemporary technoscience has thrown up an ever-wider range of boundary-crossing possibilities for both the body politic and corporeal (Franklin and Lock, 2003; Brown and Webster, 2004). Medicine is now at the heart of an array of combinatorial human, animal and mechanical materials. Whilst the troubled nature of transplantation has been relatively well documented, (e.g. Fox and Swazey, 1978; 1992), less well understood are new forms of innovation that cut across machines, humans and animals raising regulatory concerns about material and cultural risk (Brown and Michael, 2004; Faulkner et al., 2004). At the same time, such hybrids are powerful sources of hope – new treatments for large populations throughout the aging societies, and new sources of wealth for countries seeking a place in the emerging tissue economies.

Hybridity takes many forms and contemporary developments in the manipulation of tissues have extended these profoundly. Hybrids signal the breach of various socio-material categories, indicating inconsistencies that disorder routines and accepted mores. It is no accident that concepts of pollution and contamination have had an increasingly important place in sociological and anthropological accounts of the life sciences lately. These disruptions are frequently framed around questions of new standards for the purity of cell lines, the cleanliness of animal tissues, new rules to secure safety and avoid hazard.

Often falling outside existing frames of institutional and disciplinary understanding, hybrids are messy/disorderly creatures. They are 'matters in wrong places' (Douglas, 1966) and for regulation they have consequently become 'matters of concern' (Latour, 2004). This chapter compares the regulatory ordering of human tissue engineering (TE) and xenotransplantation (XT), areas of innovation which regulators have sought to govern separately and in isolation from one another. Contrasting definitional boundaries and regulatory mechanisms partition them socially and institutionally. But despite these attempts at purification, TE and XT have proven increasingly difficult to tell apart in practical and material terms. Human and animal matters, cell cultures and tissue products have much greater corporeal connection than has been institutionally recognised.

This chapter tells the story of how the messy worlds of TE and XT have leaked into one another, calling into question the abilities of regulation to adequately control hybrid innovations. Clinical XT has been subject to forceful regulatory prohibition over recent years with the relatively stable consensus that risks outweigh benefits, particularly in respect to transpecies disease. But recently the regulatory grip on XT has been called into question as various areas of TE have come to regulatory attention. This has resulted in significant attempts to 'clean up' the definition of XT, broadening its regulatory identity. 'Tissue engineering' on the other hand signifies a far more diverse set of practices whose identity in regulation has been more plastic than that of XT. As such, regulators have sought to draw a clean distinction between different aspects of implant innovation, steering TE clear of potential contamination by more problematic hybrids like XT and stem cell therapy. The chapter asks whether such distinctions can be sustained.

First the chapter sets out its conceptual terrain then discusses xenotransplantation and human tissue engineering drawing on data from two social science research projects. To conclude we consider the implications of these comparisons for understanding the variability of governance in human implant technologies.

Theorising regulatory hybridity

Novel natures – like TE and XT – are enmeshed in the production of equally novel/hybrid regulatory orders and institutions, processes whereby nature and the social are made available to each other in what Rabinow describes as 'biosociality' (Rabinow, 1992). Here we are interested in various forms of 'institutional biosociality' (Brown and Michael, 2004) through which scientific and regulatory actors – in TE and XT – configure one another materially, culturally and institutionally. In other words, regulatory bodies form particular representations of corporeal bodies and in turn subject corporeality to the innovativeness of regulation.

Hybrids present regulation with the need to alter the boundaries between existing institutional arrangements. Stem cells traverse the borders between regulated reproduction and transplantation (Waldby, 2002; Franklin, 2001).

Pharmacogenomics newly combines the regulation of genetic diagnostics and medicines (Hedgecoe and Martin, 2003). TE and XT are similarly hybrid falling into a 'regulatory vacuum' (Faulkner et al., 2003) between drugs, devices, human implants and animal research. Hybrid regulatory capacities are therefore 'risky creatures' (Brown and Michael, 2004).

Innovation occurs at the limits of conventional organisational arrangements (Gibbons et al., 1994; Nowotny et al., 2001) challenging the taken for granted and presenting novel risks. New regulatory bodies can be seen as institutional interpretations of the composition and materiality of these novel risks. Various regulatory elements are assembled together in such a way that they reflect, often imperfectly, new regulatory objects, producing intricate connections between 'natural' and 'institutional' hybrids, or as Martin puts it 'how governance arrangements are being challenged and transformed cannot be detached from these other sociotechnical changes' (2001: 158). More radically, regulatory work may also be seen as a powerful force in the very conception and conceptualisation of innovative technology (Bud, 1995: 297).

Regulation is innovative. Here we employ the concept of the 'regulatory order' (Faulkner et al., 2004) and 'regulatory ordering' to draw together strands of theory relevant to the innovativeness of hybrids. Firstly, we can regard corporeal and institutional hybridity in terms of cleanliness and dirtiness (Douglas, 1966). Pollution is a fundamental axis in the dynamics of everyday life. 'Dirt is essentially disorder' (1966: 12), 'matter out of place' which must be excluded if order is to be maintained, insights into boundary-work and categorisation that have proven highly influential in the analysis of medical and scientific knowledge (Bloor, 1978; Gabe, 1995; Carter, 1995; Mody, 2001). But arguably the Douglasian perspective, while bringing into view the value-laden partitioning of material and social boundaries, assumes relatively static categories – pollution as the expression of an underlying structuralist order.

By contrast we look to a second set of theoretical perspectives which are more dynamic and indeed even celebrate transgressive intermixing (Ansell Pearson, 1999). The two views might be articulated in terms of Deleuze and Guattari's (1988) conception of 'being and becoming'. Between the extremes lies a more ambivalent position that can be identified with, amongst others, Haraway (e.g. 1991a, 1991b; Prins, 1995). Myerson has unpacked this ambivalence in relation to Haraway's (1997) writing on the oncomouse, a profoundly ambiguous figure with whom 'we can acknowledge our kinship ... either as victims or as heroes' (Myerson, 2000: 73). Haraway's hybrid, the cyborg, also has political potential in terms of holding out the prospect of couplings that ultimately demolish oppressive dichotomies operating across genders, races, species and machines. And for Latour hybrids are perhaps even more ambivalent (1993). They are often dangerous, even catastrophic – the ozone hole, climate change, BSE – and are evidence of the underlying hidden connections between humans and nonhumans, and especially between 'objective' science and 'subjective' politics. Hybrids challenge representational order, disturbing the very basis of modernity's sorting. And of course, they can be highly hazardous unless recognised as such in political process, a 'parliament of things' (ibid). Hybrids have agency in as much as they can undermine attempts to conceal connection.

Regulatory re-ordering, therefore, requires the important concepts of cleaning/dirtying of normative categories, pollution, purification and decontamination. Equally, it requires the sense that hybridity is much more highly unstable and volatile, constantly challenging systems of classification and the material boundaries of technology.

The jurisdictional fields of socio-technology that regulation attempts to define can be usefully conceptualised as 'zones' (Barry, 2001) or 'territories' (Sharp, 2002). It is part of the regulatory construction of such zones that the technologies that we will discuss here have come to be known conventionally as 'tissue engineering' and 'xenotransplantation'. Jurisdictional structures can be difficult to establish in processes of regulatory ordering, but are 'meant to invoke order and to demarcate boundaries' (Hogle, 2002: 243). Processes of regulatory ordering engage with a fluid governance jigsaw – a web of interlinked laws, regulations, guidance and surveillance interacting with the negotiation of technology zones. Regulation exhibits, *par excellence*, societies' attempts to establish links between innovative technologies and the social management of their opportunities and risks, stabilised in institutions and patrolled through standard-setting and surveillance.

As we illustrate in the two cases that follow, the ordering of the regulatable zones of bioscience is in the view of many stakeholders – though not all – severely challenged by human and animal tissue-based technologies. In this discussion we will see that both XT and TE have been subject to important contestations over their definitional, regulatory identities – that is, what is or is not to be considered inside or outside their leaky borders. We now present these two cases as examples of the workings of 'regulatory re-ordering'.

Xenotransplantation

XT has hovered controversially on the horizons of biomedicine for decades but remains firmly locked within a whole range of presently prohibitive dangers. Nevertheless, some research has continued together with corresponding developments in regulation and public consultation. One of the foundational problems for regulation in this area has been how to define xenotransplantation separate and distinguishable from other developments in bioscience. That is, what are its limits? What is and is not XT? How might regulators 'clean up' the messy worlds of bioscience such that they have a precise understanding of their regulatory object?

In 2001 the US Food and Drug Administration revised its definition of xenotransplantation as it became aware of an apparently 'unrelated' medical procedure which had until then not been considered relevant to xenotransplantation regulation. Since 1987 Genzyme had been marketing Epicel™ a method for culturing human skin for treating severe burns. The problem was that the production method requires base layers of irradiated mouse cells on which to culture human epidermal grafts.

Xenotransplantation regulators saw in this the very same dangers that concerned them about transpecies transplants, particularly human/nonhuman disease transmission through viruses embedded deep within the DNA of a species, that, when introduced into an unrelated species can become highly infective and harmful.

Years earlier, Epicel™ had been originally regulated as a 'medical device', though its connection to the risks associated with XT were not foreseen at the time, nor included in the definition of XT that emerged during the early and mid 1990s. But with a now increased regulatory focus on transpecies disease the status of Epicel™ as a 'device' began to collapse, as did the existing definition of XT. The previous definition used by the FDA and the British regulatory body for XT (the United Kingdom Xenotransplantation Interim Regulatory Authority) did not account for production methods whereby human and animal tissues may be subject to 'ex-vivo contact' as is the case with Epicel™. Whilst they foresaw that there would be the need to regulate clinical practices involving ex-vivo 'perfusion' (pig livers used temporarily outside of the body to support a patient in hepatic failure), ex-vivo contact remained outwith the duties of responsibilities of regulators. The amended FDA definition now reads:

> ... any procedure that involves the transplantation, implantation, or infusion into a human recipient of either (a) live cells, tissues, or organs from a nonhuman animal source, or (b) human body fluids, cells, tissues or organs that have had ex vivo contact with live nonhuman animal cells, tissues or organs. (FDA, 2001)

This realisation in the late 1990s and the FDA's redefinition, undermined the adequacy of existing regulation and also the belief held by many authorities that they had produced a precautionary regulatory process. Regulators had been fundamentally committed to the principle that recipients of xenografts should be registered within a programme of life-long surveillance, monitoring potential cross-species infection. But in testimony to the FDA, representatives of Genzyme acknowledged that whilst many thousands of patients had been treated internationally with Epicel™, there was no registry of recipients and no means of checking for cross-infection. Awareness of Epicel™ and other products suggested that the xeno-horse had indeed already long since bolted.

So these questions of definition and what is or is not the object of regulatory action, and by which regulatory institution, are therefore far from trivial. And in this case there are issues about whether a new clinical practice is a device/appliance or a medicine, or more crucially whether it is alive or dead, inert or animate, benign or dangerous. As one respondent from the UK Department of Health commented:

> ... the mice who provided the cells were long dead. Years and years ago, so it's an established cell line that's being used, really without people thinking about it as being particularly mousy. (DoH member 1 – June 2001a)

Questions of species were crucial here – a device's 'mousiness' or not! Initially, the 'radar' of regulation was focused on the species that looked most likely to be the primary donor source, pigs. Therefore, characterising the pig and its potential donor-host relationship to recipient humans occupied most of the regulatory workload. Non-human primates too were at the centre of sustained discussion, particularly in respect to their welfare as proxy humans in preclinical trials. Plus the focus was initially skewed towards large organ transplants and perfusion techniques from pigs rather than cellular applications involving mice, and perhaps other slightly more mundane technologies including Epicel™.

In all, cells and mice were arguably peripheral in the regulatory consciousness:

> first it was all about ... whole organs ... concern as a committee tended to shift more towards the use of animal tissues ... Infusion devices, transplanting foetal cells, this kind of thing ... this arose out of an American experience where their definition included any tissues which had been in contact with animal body fluids, and this then became relevant in certain things ... notably the growth of tissue cells, using at one stage mouse cells as a base, and that this was not xenotransplantation according to our definition because no cells or vital animal tissues were transplanted. But human cells have been in contact with mouse cells. And this created quite a difficult problem ... because there was a possibility that we might have inadvertently created a lot of xenotransplant patients simply by changing the definition. (UKXIRA member 2, Oct 2001)

As these definitions shift so too does the ground on which regulators understand transpecies risk. As one of our respondents put it, there are differing degrees of risk between whole organs which are transplanted and irradiated mouse cells which are not. In this case there is the difficulty of balancing the fact that there are now many more various forms of xenotransplantation and differing degrees of contact, but 'contact' nonetheless. An expert witness from Genzyme at an FDA committee hearing acknowledged that contact may be sufficient cause for regulatory concern:

> I think the assumption that one takes using any mouse cell line is that there is the potential for expression of a xenotropic retrovirus from that

line. The fact that these tests are negative is comforting at the post-production level, and the fact that they're radiated prior to their use as a feeder cell line is comforting to some extent in that they're not actively replicating. But there is always a potential with any mouse line the assumption being that there is potentially endogenous retrovirus there. (D. Moore, Genzyme, FDA hearing, April, 2000)

This account shows how the once 'black boxed' (Latour, 1987) definition of xenotransplantation was opened up to controversy, how the 'zoning' of XT has expanded, and evolved in relation to institutional assumptions about species difference, and about whether something is an inert device or an active biological agent. The focus on pigs distracted the regulatory gaze from other xenogeneic hybrids including the living cells of long dead mice and the many thousands of human skins cultured on them. The broader point is that the *disjunction* between the apparent hybridity of a body like UKXIRA and its actual (species) rigidity generates apparently new risks.

In this sense then, hybrid corporealities are in profound tension with hybrid institutional bodies – what we described earlier as a form of 'institutional biosociality'. That is, institutions – like other areas of specialisation within the biosciences such as professions, disciplines, royal societies, etc. – have a certain species identity or emphasis. In this way they may be aptly described as 'institutional animals', each resonating with particular elements in nature.

In the UK, two regulatory bodies in particular emerged as crucial to the management of human and nonhuman risk in XT: UKXIRA, but also the Home Office's Animals Procedures Committee (APC). UKXIRA is an interesting illustration of a hybrid or transpecies institutional capacity. Its terms of reference include oversight of applications to undertake clinical trials in humans and also consideration of the welfare of source animals from which tissues and organs might be derived. Its remit, its routes of consultation and even its membership criss-cross various 'relevant' regulatory capacities that feed into its role as an intermediary focus in the UK regulatory system for XT:

> The authority's role as a focal point for xenotransplantation issues is important given the number of interests which xenotransplantation brings together – animal and human welfare and ethics, industry, public health, and the other regulatory systems which exists for medicines and medical devices ... (UKXIRA, 2003: 5)

The important point here is that it is, we might say, a predominantly human-medical regulatory animal. On the other hand, we also have the Home Office's APC, responsible for preclinical experimental procedures on animals, a nonhuman-welfare regulatory animal.

Whilst both of these committees are crucial regulatory bodies for XT they have been entrenched in institutional arrangements that inhibit good joint responsibility for 'interspecies' regulatory problems. A crucial issue here is that the APC is bound by clause 24 of the Animal (Scientific Procedures) Act 1986 making it a criminal offence for members of the APC to disclose information to third parties, 'otherwise than for the purpose of discharging his functions under the Act'. This also applies to the disclosure of information to other regulatory authorities, even when their terms of reference may directly apply. One of our respondents sat on both of these committees and found themselves at the centre of critical institutional tensions.

> The APC ... is all tied up with confidentiality ... when somebody applies to the APC to do a project on xenotransplantation involving primates, the logical thing to do would be to refer back to the body that regulates xenotransplantation [UKXIRA] ... but you can't do that because UKXIRA cannot have sight of the project licence application ... (UKXIRA/APC member 1, April 2002)

Another respondent was a member of UKXIRA but not the APC. They expressed their frustration at being consulted by the APC about licence applications, the contents of which they were forbidden from seeing:

> ... the APC ... doesn't allow the dissemination of information to anybody outside the committee. I mean the sort of issues which arise are whether a series of experiments on a group of primates would be of sufficient importance ... to justify the suffering ... And we would get a question about the health of the animals from the Home Office but we couldn't get details of the experiments, and we couldn't get details of previous experiments to put them in context ... (UKXIRA member 2, Oct 2001)

There is then a legal firewall here in the governance of humans and animals, reflected in the structure of government departments and regulatory bodies. These species divisions represent potential weaknesses in the risk management of innovations like this that cut across institutions and natures. So despite its hybrid character, UKXIRA clearly has a species identity. Its institutional location reflects a stronger alignment with networks of human-medical-governance than it does with those of animal-welfare governance. Increasingly, clause 24 has come under greater scrutiny but is likely to remain in place until legal process has taken its long and circuitous institutional course through processes of consultation, lobbying and parliamentary timetabling. In the mean time, hybrid bodies like UKXIRA remain pragmatic and rely on a variety of strategies with which to extend their hybridity and to better meet their terms of reference (UKXIRA, 2003: 9).

This story illustrates the limits to institutional hybridity, that the capacity of regulatory bodies to move smoothly across long established institutional structures is limited. More importantly, this reflects institutional representations of boundaries that are seen to exist in nature, and of course those boundaries traversed or innovated in biotechnology. Xenografts, as with other biotechnological innovations, often belie both 'natural' and 'institutional' classification, as we could see in the debates about the definition of XT, challenging the way in which routes of responsibility are organised. But as we will see next in our discussion of tissue engineering, these problems of 'making a mess' and 'cleaning up' are endemic as novel natures are generated and the definitional and institutional alignments of regulation become unsettled.

Tissue engineering

The discussion of XT highlights a number of key features of regulatory re-ordering that are important to consideration of human tissues and TE including the purification of the regulatory object, a strong alignment of regulatory institutions with one rather than another substantive field of governance, and the dynamic tension between institutional hybrids and novel natures. We now draw upon recent research – principally in the UK and Europe – elaborating TE as a regulatable zone and in relation to XT.

More heterogeneous than XT, TE includes cultured cell implants for cartilage repair, bone substitutes and 'living' skin tissues (like Epicel discussed above). Future developments are expected to include vascular prostheses, organ-assist devices (liver, kidney), whole organs, structures (heart valves, joints), neurological tissues and stem cell therapies. One influential definition describes TE as the 'regeneration of biological tissue through the use of cells, with the aid of supporting structures and/or biomolecules' (SCMPMD, 2001a). They are often conceived of in regulatory policy communities themselves as 'borderline' or 'hybrid' products – occupying a 'regulatory vacuum' at the borders of existing regulatory frameworks. The notion of a regulatory vacuum has been the starting-point for much of the recent development of regulatory policy for human tissues, and the boundaries between pharmaceuticals, medical devices and TE are crucial areas of negotiation in re-ordering regulation. Some TE products have already been regulated as pharmaceuticals whilst some parts of combination products (e.g. using synthetic scaffolds) need to gain approval as medical devices. In the face of this complexity there has been a widely, though not unanimously, perceived need for 'new regulation' for human tissues and TE.

TE embraces two closely related regulatory fields whose boundaries are themselves unclear and overlapping. On the one hand, 'human tissues and cells' (HTCs) are defined and covered in Europe by the Tissue and Cells Directive (TCD) on the sourcing, storage, processing (e.g. cleaning) and distribution of a wide range of human materials (excluding, as we shall see, matters like blood and blood products and whole organs). On the other hand, the proposed 'human tissue engineering regulation' (TER) refers to manufacturing and market approval, excluding the accreditation of safety and quality of sourcing and storage covered by the TCD, and distinguishes between autologous (donor is also patient) and allogeneic (multiple recipient) applications. These two jurisdictions were already separated in the United Kingdom, whose regulatory work in the early 2000s included a tightening of standards and accountability through a code of practice for tissue banking (Department of Health, 2001b), and then a code of practice for 'human-derived therapeutic products', voluntary guidance for manufacturers in the TE field (MDA, 2002).

Importantly, the European directive concerned with sourcing and storage is nevertheless framed, as discussed below, in terms of transplantation and therapeutic application. A TE manufacturer or tissue establishment engaged in significant tissue manipulation would have to meet requirements of both fields. These ambiguities are characteristic of the hybrid aspects of tissue-based therapies, pointing to a distinction between the traditional tissue banking for transplantation and the emerging activity of engineering tissues in implants. This is an increasingly troubled distinction as tissue banks engage in manipulation, and industry engages in tissue 'banking'.

It has been important for both commercial and regulatory interests that a clearly delineated zone of activity can be identified. European Commission Directorate-General Sanco negotiated the TCD in seeking to fulfil its requirements to protect public health, prefacing this directive that 'The transplantation of human tissues and cells is a strongly expanding field of medicine offering great opportunities for the treatment of as yet incurable diseases' and 'As tissue and cell therapy is a field in which an intensive worldwide exchange is taking place, it is desirable to have worldwide standards.' Note here the inclusion of 'cell therapy' in their definition.

Biomedical zones are matters of negotiation, with national and sectoral interests playing an important role. Unlike XT it is not the case with TE and HTCs that an initially stabilised definition has been changed by a later revision, but that a range of definitions have been proposed and continue to be negotiated, taking the form of partitioning and cleaning with attempts to distinguish TE and HTCs from xenografts. In what follows we also illustrate further attempts to cleanse TE, variously including or excluding embryonic stem cells, whole organs and blood products.

We saw how the definition of XT had been formed in isolation from regulated 'devices', and how, given the production method used (irradiated cultures of nonhuman cells) that zone began to collapse, overwhelmed by the hybrid linkages. Now, whilst XT regulators have been slowly expanding the definition of XT, just the opposite has been taking place in human TE. Regulatory efforts have been directed, with mixed success, at distancing TE and human tissues/cells from the allied worlds of XT. One respondent spoke of how regulators sought to maintain a firm distinction between TE/HTCs and XT:

Q: What about the use of animal or … ?
A: Animal? for me it's a separate area … xenografting … we should have a centralised authorisation system for that.
Q: if you look at Apligraf [TE wound treatment] … it makes use of bovine serum during the culturing process.
A: Oh, we have a problem of definition here again because for us a xenograft is the use of the animal part in the body or in the perfusion extra-corporeal. But it's a direct use. We have a specific category of products we call Produits Thérapeutique Annexe which means additives you could say. And we have a specific authorisation for additives.
Q: Including?
A: … bovine serum in many culture processes. … And our regulation of xenograft existed since '96 … that only the Health Minister can authorise a xenograft trial … You can see that it's been placed at the most important level. (A-EU4, 2003)

Essentially, however, the respondent is describing a national regulatory boundary TE and XT that no longer exists in the US and will cease to exist in those regulatory arenas that take their lead, like the UK. The distinction of two regulatory jurisdictions will no longer be tenable, with both defined as XT. This clearly signifies the increasing reach of cleansing policies, with 'problems of definition' potentially jeopardising both the political and material cleanliness of TE and HTCs. Critical clinical commentary also recognises this:

> In ex-vivo corneal stem cell expansion … there is a need for accreditation of laboratories conducting such work. The use of … of co-culture system must be in consultation [with regulators]. Without such stringency, there will be a risk of cross-animal contamination such as the one we witness [*sic*] recently in outbreaks of SARS. (Kong Y. Then, 2003)

This writer recognises the extension of XT and the role of the TCD in providing a framework for international accreditation. These processes of cleaning and re-organising are as much material as they are institutional and regulatory, calling to mind what Rheinberger describes as the 'intracellular representation of extracellular projects' (2000: 19). The quotation further illustrates the ambiguous boundaries around HTCs and TE, where adult stem cell therapy *may* be regarded as a form of TE.

In what follows, a respondent in the academic-commercial TE sector talks about their attempt to materially re-engineer the methods used to culture skin cells, seeking to replace the use of animal 'feeder layers' with human equivalents, reflecting similar attempts across the TE industries.

> We are very keen to develop a methodology that doesn't use any bovine materials. Personally I'm more concerned about using bovine than mouse cells as a feeder layer. ... [XT regulators in the UK] are very concerned with our groups using mouse fibroblasts, but when I ask them 'aren't you concerned about bovine material?', they say that's not part of our remit because the cells are not alive. At which point you put your head in your hands and cry. We are working very hard to ... [develop] a methodology for using the patients' own fibroblasts to substitute for bovine serum. ...(S5, 2003)

The statement highlights a further classificatory distinction in TE regulation, that of viability/non-viability. Here we see a hybrid 'consumer' of regulation troubled by the codified distinction between XT and medical device, and the regulatory vacuum for human TE products. (European medical *device* directives cover the use of *nonviable* animal materials in medical device manufacture). Industry has sought to maintain the limited relevance to final TE products of 'ancillary' *viable* XT in production processes, as we will see below. The following statement from the European parliamentary debate on the draft TCD concedes some serious ambivalence about the exclusion of XT in the new legislation for HTCs:

> Organs, tissues and cells of animal origin for human therapy are still in the research phase, but nevertheless pose different regulatory problems that will need to be addressed in due course (from the Explanatory memorandum to the proposal for a directive on quality and safety of human tissues and cells (EU Commission/DG Sanco (2003))

Animal tissues and cells were in fact excluded from the final Directive. The EC here is thus acting to preserve the integrity of HTCs as a regulatable zone uncontaminated by animal matters.

The EC's DG Enterprise found in its consultations on TER – the TE-specific regulation in 2002 and 2004 – that most stakeholders favoured a separate regulatory framework for XT *products*. However, some attempt to address the question about contact with animal materials during production was suggested. Responding, industry associations proposed the following text:

> hTEPs containing not intentionally small quantities or traces of material of animal origin (used during the manufacturing process) which do not perform any function *in the finished product* are not, for the purpose of this regulation, regarded as xenogenic products. (EuropaBio et al., 2004; author italics)

Thus they attempt to preserve TE production processes *against* the incursion of the extended definition of XT as discussed above in the case of the US and UK. Such *products* should be regarded as TE products in spite of viable XT elements in the production process.

The UK has a dedicated XT regulatory body but the European Union does not. The UK's code of practice for human derived therapeutic products (MDA, 2002) is the basis of interim guidance for manufacturers of TE technologies while the EU TER is being formulated. The UK code states that 'Where cell culturing techniques use cell lines that are not of human origin, for example, murine fibroblasts used for co-culture, guidance should be sought from UKXIRA.' The code specifies appropriate measures to avoid material contamination and to provide for regulatory accountability:

> Documentation shall be obtained that demonstrates the application of appropriate quality assurance measures by suppliers of biological material, including origins and veterinary certificates for the animals used in the preparation of the material (e.g. bovine serum albumin).

and:

> Culture media, reagents and processing materials derived from animals shall be evaluated for the risk of contamination with micro-organisms, particularly viruses and agents of transmissible spongiform encephalopathies ...

Thus it is clear the UK TE policy group assumed that the UK system, including UKXIRA, should and will adopt the extended definition of XT, and allows for a linkage between regulatory zones with parallels to that seen above between medical and animal welfare domains.

Returning to the clean ordering of HTCs, whole organs were also excluded from the Tissues & Cells Directive, in the face of strong dissension. As one commissioner noted:

> ... I remain convinced that it is not appropriate to include organs in the scope of this directive. The problems to solve in this area are quite different ... requiring a different policy approach ... As organ transplantation is a highly specialised subject in its own right, the Commission is currently conducting a scientific evaluation of the available options ... Following the example of the blood directive and this proposal on tissues and cells, *we would like to get the science right first, before tabling a legal instrument in this sensitive area.* (Byrne, EC, 2003) (author italics)

We see here 'getting the science right' as a rhetorical device with the effect of proliferating separate though linked regulatory jurisdictions – blood, organs, and human tissues. These distinctions between biosocial matters are artefacts of regulatory political process. But paradoxically purifications aimed at defining discrete fields at the same time increase the overall complexity. On the exclusion of organs again:

> We should not include organs in this measure on cells and tissues. Organs are for another day. Equally, this is not the time to permit cloned human embryos or hybrid human–animal embryos ... this is a very young area of science ... leaving aside the ethical issues, one that should not be permitted now. (MEP Bowis, 2003).

This remark nicely points to the complexity of the human tissue terrain and the inevitable difficulties of policing its fragile borders. It is worth remembering here the way the regulation of XT had 'mistakenly' focused on whole organs, for years neglecting cell-based 'xeno-like' practices in tissue engineering. There is then a curious patterning in the regulation of tissues, human and xenogeneic. Whilst HTC regulators would like not to have to embrace whole organs, XT regulators would rather not have had to embrace TE. The fact that cells (particularly *ex vivo* contact) are now more central in the regulation of XT poses yet more problems for HTCs and TE legislation. It is possible that the exclusion of whole organs raises similar questions about the longer-term tenability of the TCD Directive's boundaries.

Thus the leakiness of distinctions between types of human tissue and their methods of 'production' means that the isolation or segregation of particular zones of a regulatory order is difficult to achieve. Hybrids have the potential to overwhelm purification because rhetorical and material connections constantly reference new associations.

Turning to consider *institutional* hybridity accompanying the contestation of biomedical boundaries, we note that in the United Kingdom the current regulatory authority the Medicines and Healthcare products Regulatory Agency (MHRA) was formed from a merger of the Medicines Control Agency (MCA) and the Medical Devices Agency (MDA) avowedly as a response to the increase in materially hybrid or combination products. Products deemed 'tissue-engineered' are assessed in the MHRA on a case-by-case basis in the absence of TE-specific regulation. In a subsequent recent development the UK has established a dedicated Human Tissue Authority which will shortly assume responsibility for oversight of all tissue establishments and their sourcing and related activities.

Various models for an institutional regulatory agency have been proposed in Europe, to satisfy conflicting interests in the TE zone. The high status Scientific Committee for Medicinal Products and Medical Devices (SCMPMD, 2001a) proposed a separate TE Regulatory Authority. Opinion then appeared to favour founding a structure located within Europe's existing Agency for the Evaluation of Medicinal Products (EMEA). This drew criticism from those stakeholders who preferred to see TE products as 'more devicey' (as one informant put it) than pharmaceutical. The hybridity of the technology was further highlighted by apparent territorial disputes within the EC's DG Enterprise between the medicines and medical devices jurisdictions, which in a further twist resulted in a proposal to re-classify TE within the 'biotechnology' division of EMEA – neither pharmaceutical nor device! Following a Europe-wide guidance document (CPMP, 2001) in 2003 the EC Medicinal Product Directive (MPD) was augmented by an Annex on 'advanced therapy medicinal products' (EU Commission, 2003) which in effect extended the definition of the regulatory field of medicines (as opposed to devices, or biologics, or TE) to include somatic cell therapy products, human and xenogeneic.

The overlap of a future European TE legislation with the MPD definition of cell therapy medicinal products, and the possibility that some products would be *both* TE and medicinal, had been criticised in responses to the EU DG Enterprise consultation on the need for TE-specific regulation (2002). And it is at this point that we can return to a consideration of the chequered regulatory history and ambiguous regulatory identity of Epicel, currently described by its manufacturer Genzyme as an 'autologous cell therapy product' that is 'co-cultured with mouse cells to form cultured epidermal autografts', and uses 'a cell culture medium containing bovine serum'. Thus were Epicel and allied products to be submitted for authorisation in Europe now their regulatory status may be unclear, in spite of meeting the main criteria of TE, and even though in the case of Epicel it is regulated as a device in the US.

To summarise, the links between HTCs, TE products and the institutional hybridity of regulatory authorities in the UK and Europe are highly complex. Like XT, we see attempts to construct and align pure regulatable fields across the hybrid materiality of human tissue-derived therapies in order to control various politico-material risks. Also like XT, we see these attempts at partitioning undermined by changes in the sociotechnical definition of regulatable therapeutic materials. The XT/device divide could not be sustained, nor could the cell therapy/medicines divide, nor could the TE/medicines/devices divides as overarching distinctions. In an admission of the difficulty of assigning stable classifications to capture novel TE therapies, EC proposals for TE Regulation allow for a *lex specialis* function to adjudicate on products that are not clearly either TE or medicinal or medical device, 'to minimise the risk of grey areas for borderline products' (EC DG Enterprise, 2004). So we also see tensions between different organisational and indeed ontological claims at the very heart of the social constitution of human therapeutic materials.

Conclusion – boundaries redrawn

Unmistakable connections cut across the regulatory and material practice of life science innovation generally, particularly evident in the context of the two innovative areas discussed here. That is, engineered human and nonhuman tissues are inherently messy and liable to 'leakiness' (Hogle, 2002). Their edges, their boundaries, are for regulators annoyingly variegated, and a source of frustration in their attempts at definition, cleanliness and purity. TE/HTCs and XT both illustrate new capacities of isolation and mobilisation in life science innovation. Here, processes of 'purification' render matters isolatable, manipulable, and legible in laboratory-based science (Knorr-Cetina, 1999: 27). Cells, tissues and bodies, as Waldby (2002) points out, are increasingly caught up in 'biotechnical fragmentation' (239). Crucially, at stake here are regulatory processes of 'territoriality' or 'political ecology' (Sharp, 2002). That is to say, natural objects are delegated to various arms of regulatory order, institutionally enacted readings of biological risk that subsequently order cells, tissues, embryos.

There are important lessons to be learnt from hybrids and dirt. Crucially, mess is a consequence of purification and not a cause, a 'by-product' of ordering and for Latour, it is the very act of purification that proliferates the production of hybrids. Boundary-making is intended to deny connection, to foreclose the production of hybrids, and so paradoxically acts to facilitate their manufacture. And all too often, regulatory ordering systematically obscures the complex interplay of regulated matters. Risks flourish, it seems, when practices of regulatory purification continue to be applied, in ignorance and denial of evident associations between technical and social considerations.

This prompts crucial questions of regulatory activity in the areas of HTCs and TE and XT – especially the sustainability of regulating them separately when based upon political or commercially pragmatic differences. This is not to suggest that acts of purification and cleaning-up are bad, even avoidable. They are not, especially in a context where transpecies innovations depend upon strong and rigorous regulatory ordering to lessen the chances of potentially devastating population-wide risks. As Barad puts it, 'boundaries are not our enemies' and we can hardly expect to do without them, they are:

> ... necessary for making meanings, but this does not make them innocent. Boundaries have real material consequences ... Our goal should not be to find less false boundaries for all spacetime, but reliable, accountable, located temporary boundaries, which we should anticipate will quickly close in against us. (Barad, 1998: 187)

Biotechnological innovations are reciprocally enabled by regulatory structures that facilitate particular sorts of research regimes and interactions out of which emerge biotechnological innovation – what we describe above as a form of 'institutional biosociality'. These are highly heterogeneous interactions between multiple forms of social, biological and institutional participants who jointly constitute innovation (Callon et al., 1986; Bijker,

1995; Hughes, 1983). Indeed, it might be useful to think of this as an institutional form of 'intercorporeality' (Waldby, 2002; Weiss, 2000), the connections of identification and disidentification between bodies that are as 'inter-institutional' as they are inter-embodied. That is, various innovated corporealities (stem cells, growth media, pigs, mice, primates, plants, viruses, patients, etc.) are distributed between regulatory bodies each participating in a complex process of exchange and interaction, potentially embodied in both donors and recipients of transplantable tissues.

In summary the messy material hybridity of biomedical regulatory objects highlights societal attempts to introduce clear partitions, jurisdictions and stable regulatory orders in highly complex socio-political zones. We have illustrated how the formations of XT and TE have interacted with shifting regulatory terrains. We have seen both successful and unsuccessful attempts to maintain boundaries, particularly between XT and human tissues. In both fields we have seen the strength of biomedical and pharmaceutical 'institutional animals' in shaping the discourse in which novel technological governance is constructed. And we have seen the shifting definitions of scientific and social appraisal of public health risk being refracted through the composition of the regulatory bodies that governance activity produces. Thus in order to understand the variability of governance in different innovative technological fields, it is necessary to develop accounts of the hybridity of their material forms, to bring into view the detailed social, cultural and material shaping that produce regulatory orderings. And it is here perhaps that we can return to the link between regulatory ordering and deeper categorising of nature, the human, animal and the moral. For the messy work of regulating is also an inalienably social and moral process of seeking benefit and minimising risk in highly pluralistic, technologically inventive societies. Hybrid technologies highlight the manufactured disturbance of foundational categories and societies' attempts to manage this disruption, a partitioning and aligning process that, in turn, redistributes the productive elements for the continuing hybridisation of biomedical technologies.

Acknowledgements

The findings presented here draw on two ESRC supported projects: 'Medical Device Governance: Regulation of Tissue Engineering' (L218 25 2058) and 'Xenotransplantation: Risk Identities and the Human/nonhuman Interface' (L218 25 2044). The authors would like to thank in particular the work of Ingrid Geesink for her contribution of the research on tissue engineering.

14
Cultural Politics and Human Embryonic Stem Cell Science

Brian Salter

Introduction

The emergence of any new health technology occurs within a cultural context that may be more or less sympathetic to the technology. On the whole, we can expect that the values of the science that creates the technology and the industry that markets it will be supportive. For example, the values of 'scientific freedom' and 'enhancing the wealth of the nation' provide the basis for broad legitimations of health technologies. However, the response of the values of civil society to new health technologies is less predictable. Some health technologies, particularly those following a familiar and well worn path of development, will stimulate little opposition that cannot be dealt with by asserting the value of their projected health gain. Others, particularly those employing innovative forms of science, may engage with cultural values of civil society that are resistant to their development and use. Such engagement can result not only in an erosion of public trust in the new technology but active opposition to its introduction. In this situation, as the example of GM food and agriculture ably illustrates, consumer confidence is undermined and potential markets threatened.

It is the promise of the scientists engaged in human embryonic stem cell (ESC) research that their work will lead to therapies capable of dealing with one of the major challenges of modern medicine: irreversible organ and tissue failure. Their claim is that research on human embryos can enable the production of 'pluripotent' stem cells, undifferentiated cells that have the capacity to develop into almost all of the body's tissue types, and thus provide the ability to create an unlimited supply of transplantable tissues. Whole organ transplants, such as for heart or kidney disease, will no longer rely on a supply of donors, neurodegenerative disorders such as Parkinson's and Alzeimer's disease will become treatable, and chronic conditions such as diabetes will be given new tissues capable of replacing the function of pharmaceutical regimes. Tissue engineering and regenerative medicine, it is argued, will be revolutionised.

This chapter examines the politics of the interaction between the immense therapeutic promise of human ESC science and the cultural values of the Member States of the European Union (EU) within the political arenas of the European Commission, European Parliament and Council of Ministers. It deals, firstly, with the factors that influence and shape the translation of a novel scientific development into a high profile political issue. Secondly, it explores the major components of the cultural politics surrounding the debate about human ESC science and Framework Programme Six: the position of the Member States, the role of bioethics and the nature of cultural trading between different positions. Thirdly, it places cultural politics in the context of institutional struggle between the Commission, Parliament and the Council of Ministers, showing the mobility of supporters and opponents of human ESC science and the range of cultural compromises they sought to achieve.

The politics of cultural mobilisation

Human embryonic stem cell science is problematic because it is dependent on the manipulation of what for many is an important part of their cultural identity: the human embryo. Human embryonic stem cells are derived from the inner cell mass of five to six day old pre-implantation embryos called blastocysts. In order to derive ESCs, the outer membrane of the blastocyst is punctured and the inner cell mass with its pluripotent stem cells transferred to a petri dish containing a culture medium. The blastocyst is thus destroyed and the ESCs can be cultivated *in vitro* to produce a stem cell line. Stem cell lines can then be modified so that they differentiate into the cells of particular tissues depending on the therapeutic objective of the investigation.

It is thus the moral status of the human embryo that constitutes the focus of the cultural politics surrounding human ESC science. The debate about the value of the embryo and its manipulation and extinction is not, of course, a new one. Over the past three decades, the issues of abortion and *in vitro* fertilisation (IVF) have produced their own kaleidoscope of different national responses (see e.g. Mulkay, 1997; Green, 2001). What is new about human ESC science in terms of its political significance is that its global economic potential for regenerative medicine and the perceived national and regional advantages to be gained from an early commitment to the field take that debate into a different league of international competition. And with the increased competition comes considerably enhanced pressures from science and industry for a political mechanism that can deal with the cultural issues stimulated by the new science.

As the Special Eurobarometer 2005 survey of 'Social Values, Science and Technology' demonstrates, whilst Member State citizens are generally supportive of medical technologies they vary widely in their views on health technologies based on innovative scientific techniques such as cloning human embryos to make cells and organs that can be transplanted into people with diseases – a central feature of human ESC science (European Commission, 2005: 85–6). Given these differences, and the different national constituencies they represent for the politicians of the EU, the question then arises of what factors influence the course of their mobilisation into competing policy discourses. In his work on stem cell policies in the United States and Germany, Gottweis has stressed the importance of 'the discursive construction of spaces of decision-making in the scientific-political field of human embryonic stem cell research'. He continues:

> The meaning of a part of the human body, such as a stem cell, or the relevance of embryonic stem cells for medical research is the outcome of a struggle between competing language games or discourses, which transform 'what is out there' into something that is socially and politically relevant. Actors mobilise what I have called heterogeneous systems of representation, intermediations between the symbolic and the material, that give broadly accepted, hegemonic definitions of the political and scientific realities, rationalities and types of actors and institutions involved in a policy field. (Gottweis, 2002: 465)

Gottweis shows how this process of politicisation occurred in the United States and Germany. At the EU, the process was condensed and intensified within just three linked institutions. As the mobilisation of cultural values took place, the initial question was the most important one: who would control the policy agenda? As Wolpe and McGee observe of the stem cell debate in the United States, 'Public policy debates are exercises in rhetoric. The first battle is often a struggle over definitions, and the winning side is usually the one most able to capture rhetorical primacy by have its definitions of the situation accepted as the taken-for-granted landscape on which the rest of the game must be staged' (Wolpe and McGee, 2001: 185). Once established, such definitions then determine which policy options are on the table and which excluded. An important tactical consideration is how the media is recruited to support one position rather than another. In the United States, Wolpe and McGee claim that guided by interested scientists and ethicists, 'the scientific media steered the debate carefully down the middle road and defined hES cells out of all problematic categories: they were not embryos, they were not cloned cells, they were not totipotent' (Wolpe and McGee, 2001: 192). Such a one-sided form of policy discourse with one set of cultural values securely in the ascendancy was always unlikely to be a feature of the EU's treatment of the issue.

The mobilisation of human ESCs as an issue where cultural values play a significant role does not, of course, take place within a governance vacuum. Traditionally, the governance of innovative health technologies has been dominated by the technocratic approach to regulation where apparently calculable quotients of such measures as costs, benefits and risks are formulaically presented within an organising theme such as the 'precautionary principle' and mediation takes place through the exercise of scientific authority (see e.g. Jasanoff, 1995). However, since for some participants in the human ESC policy domain scientific authority is no longer regarded as a legitimate basis for policy decisions, cultural conflict requires the introduction of a different authority, or authorities, to enable compromises to be reached and policies produced. Broadly speaking, two innovations in the policy process have so far been pursued: the use of an enhanced democratic input and expert bioethical advice.

In reflecting on the problem faced by policy makers, Weale observes that

> constructing successful public policies to deal with complex technologies requires us to acknowledge both the necessity of the distinctive understanding that comes with science and the need to supplement that understanding with a responsiveness to the democratic discussion of the implications of scientific analysis in the context of public values. (Weale, 2001: 415)

He then discusses the relative merits of a range of possible participative techniques (focus groups, consensus conferences, deliberative polling) for improving the democratic responsiveness of biotechnology policy making. However, although these techniques may expand the opportunities for the public to express their values, the problem of how to reconcile values that may be deontological (absolute statements of right and wrong) remains.

In this context it is increasingly the bioethicists who claim that formal ethical debate can be used as the means for noting, organising and pronouncing on the cultural arguments stimulated by new science and technologies, particularly where human genetics and the human embryo are involved (Salter and Jones, 2002). National and international bioethics committees such as UNESCO's International Bioethics Committee and the European Union's European Group on Ethics have assumed a growing importance in the global politics of sensitive new technologies. By providing a framework within which often emotive cultural divisions can be addressed, bioethics committees may enable disputes to be defined in ways capable of producing policy outcomes. That is to say, they seek to promote a form of discourse about values that is ultilitarian in its policy ambitions rather than simply a forum for the exchange of views. Provided that the participants accept the legitimacy of the process (which not all do), bioethics committees can facilitate the negotiation and exchange of cultural values and the recognition that such values are political resources that may be traded or retained. They do so by translating the values of others into what they claim is the 'neutral' discourse of bioethics and so circumvent the deontological problem (Evans, 2002: 8).

Cultural politics in Framework Programme Six

The need for a political mechanism such as bioethics capable of dealing with cultural divisions has been particularly acute in the case of ESC research and the EU's Framework Programme Six (FP6) 2001–06. The inclusion of ESC research into the FP6 agenda brought together into a single political stream two dominant prerequisites of a successful European R&D policy: the funding of basic research and the regulatory framework necessary to make that research politically sustainable across the Member States. Given the cultural politics of the human embryo, the latter was always going to be a tall order.

The propulsive forces of scientific freedom and industrial ambition had encountered substantial cultural opposition to the plans for human embryonic stem cell research from within the institutions of the EU well before FP6 was launched in 2002. In general, the conflict between the cultures of science and industry, on the one hand, and some parts of civil society, on the other, was formalised (and to a degree normalised) through the referencing of 'ethics' as a suitable and legitimate vehicle for the conduct of the continuing political bargaining. Ethical debate became the politically acceptable face of cultural conflict in that it facilitated a way of noting and organising differences between often emotive positions.

Thus when, in the formulation of the research agenda of FP5, DG Research faced a clash between the demands of science for human embryo research and the cultural opposition of certain Member States it turned to the European Group on Ethics in Science and New Technologies (EGE) for a solution. (Since 1991 the EGE has provided advice to President of the Commission on the values that should guide regulatory decisions on biotechnology.) Its 1998 Opinion *Ethical aspects of research involving the use of human embryos in the context of the 5th Framework Programme* resulted in the FP5 1998–2002 research agenda excluding any 'research activity which modifies or is intended to modify the genetic heritage of human beings by alteration of germ cells' and any 'research activity understood in the sense of the term "cloning", with the aim of replacing a germ or embryo cell nucleus with that of the cell of any individual, a cell from an embryo or a cell coming from a later stage of development to the human embryo' (European Council, 1999). Furthermore, and in the context of the EGE recommendation that FP5 should introduce the principle of ethical review for sensitive projects, Article 7 of the Council and Parliament Decision approving FP5 stated that 'All research activities conducted pursuant to the fifth framework programme shall be carried out in compliance with fundamental ethical principles' (European Parliament and European Council, 1999). In the same year, both Directive 98/44/EC on the legal protection of biotechnological inventions and Directive 98/79/EC on *in vitro* medical devices emphasised the important role of ethics in their implementation. In addition, Article 7 of the 1998 Patent Directive charged the European Group on Ethics with the evaluation of the ethical aspects of

Table 14.1: Regulations in EU Member States regarding human embryonic stem cell research (March 2003)

Type of regulatory control	Austria	Belgium	Denmark	Germany	Spain	Finland	France	Greece	Ireland	Italy	Luxembourg	Netherlands	Portugal	Sweden	UK
Prohibition of human embryo research									X						
Prohibition of the procurement of ESCs from human embryos					X		X		X						
Prohibition of procurement of ESCs from human embryos but allowing by law for importation of human ESCs	X		X	X											
Allowing for the procurement of human ESCs from supernumerary embryos by law						X		X				X		X	X
Prohibition of the creation of human embryos for research purposes by law or by ratification of the Council of Europe's Convention on Human Rights and Biomedicine	X		X	X	X	X	X	X	X			X	X	X	
Allowing for the creation of human embryos for ESC procurement by law															X
No specific legislation regarding human embryo research		X								X	X		X		

Source: European Commission, 2003: table 1 (amended).

biotechnology in general. As a formal part of the institutional discourse, ethics, *qua* ethics, had arrived. The regulation of science funding and research was no longer a purely technocratic preserve but territory where, it would seem, cultural factors had a legitimate presence.

However, although recognised as a useful vehicle for the formulation of general cultural propositions, ethics had yet to establish itself as a mechanism for the negotiation of cultural differences within the political narrative. Its political utility was therefore limited to one of presentation rather than one of resolution. Given the wide variations between Member States regarding both the definition of the human embryo and the conditions under which research, if any, could be conducted on it, as a political form ethics would have to evolve if it was to perform any kind of brokerage function between opposing cultural positions. Table 14.1 illustrates how the political cultures of Member States have found expression in a patchwork of legal arrangements containing numerous shadings of the value of, and the protection to be afforded to, the human embryo. These differences were to form the basis of the conflicts over human ESCs and FP6.

The EU debate about ESC research and FP6 constituted a struggle for control of the political narrative and thus of the policy agenda. It can be analysed in terms of a framework based on salient ethical components within the narrative derived from the moral status attached to the building blocks of human ESC science. These are therefore simultaneously scientific and ethical objects and, as the politics of the field develops, constitute the primary units of cultural exchange and trading between opposing positions (Figure 14.1). Through the use of ethical arguments for or against the use of particular components, or combinations of components, choices are made about the preferred future path of ESC science.

These components emerge both singly and in combination. Political actors would declare themselves in support of, or opposition to, the research criteria implicit within each ethical component. Figure 14.2 outlines some (but not all) of the main combinations and provides an indication of the ways in which the trading of cultural values through the medium of ethical components may be refined and conducted. Each ethical cell in the matrices can be regarded as potential agenda-setting territory and thus as a political resource that may be exchanged for, or coupled with, other cells as trading takes place within the political narrative.

- embryo source
    - aborted
    - IVF supernumerary
    - non-IVF donated
    - cloned
- embryo creation date
- embryo age
- ESC line origin
- ESC line creation date
- ESC line research purpose

Figure 14.1: Bioethics and the units of cultural trading

218 New Technologies in Health Care

At one end of the continuum of political cultures, the UK's Human Fertilisation and Embryology (Research Purposes) Regulation 2001 extends the Human Fertilisation and Embryology Act 1990 to permit the use of embryos, regardless of source, in research to increase knowledge about serious diseases and their treatment (Figure 14.2, all cells). At the other, the Irish constitution of 1937 (as amended in 1983) provides that 'the State acknowledges the right to life of the unborn and, with due regard to the equal right to life of the mother, guarantees in its laws to respect, and as far as practicable, by its laws to defend and vindicate that right' (European Commission,

ESC Conditional date of creation	Embryo Conditional date of creation	
	No	Yes
	Donated embryo	
No	1	2
Yes	3	4
	Supernumerary embryo	
No	5	6
Yes	7	8
	Aborted embryo	
No	9	10
Yes	11	12
	Cloned embryo	
No	13	14
Yes	15	16

Figure 14.2: Major ethical components of the political narrative

2003b: 42). As Table 14.1 shows, in between there lie a variety of Member State positions and non-positions constructed by states seeking to reconcile conflicting cultural pressures from civil society, science and industry. For example, in an attempt to remove the human embryo from the political equation (and/or to distance themselves from the act of embryo destruction necessary for ESC creation), Germany, Austria and Denmark have allowed the importation of ESC lines whilst internally prohibiting their procurement from human embryos. In addition, Germany's Stem Cell Act 2002 requires that the ESCs were derived from supernumerary embryos before 1 January 2002 in the country of origin (European Commission, 2003b: 40; Figure 14.2, cells 7–8). (The date of ESC line creation is, of course, the same criterion as that used by President Bush when he announced his decision to allow Federal funds to be used for research on existing – i.e. pre- 9August 2001 – human ESC lines: an example, perhaps, of transnational policy learning.)

Cultural politics and institutional struggle

The cultural conflict within the FP6 debate between, on the one hand, the moral status of the human embryo and its numerous sub-categories, and, on the other, the principles of scientific freedom and economic advance, was channelled through the institutions of the EU: the Commission, Council and Parliament. Characterised by a continuing volatility, refinements of position, and the formalisation of the role of bioethics in the institutional processes, the FP6 debate vividly illustrates the need for fresh modes of international governance to deal with cultural divisions.

Given the diversity of Member State political cultures on the human embryo, it was to be expected that the UK's report advocating greater freedoms for ESC research would prove provocative and ethically challenging. Responding to that report, on 7 September 2000 the European Parliament passed a resolution opposed to both reproductive and therapeutic cloning. Therapeutic cloning (Figure 14.2, cells 13–16) was seen as 'irreversibly crossing a boundary in research norms' and as contrary to public policy as adopted by the European Union (European Parliament, 2000). Much of the debate was couched in emotive and categorical terms with little suggestion from the opponents of ESC research that negotiation was either possible or proper. In its report *Ethical aspects of human stem cell research and use* published two months later, the EGE took a more sophisticated view and began the process of establishing an ethical continuum of types of human embryo and ESC research, using the kinds of criteria employed in Figure 14.2, and suggesting that some criteria are more acceptable than others (EGE, 2000). While it regarded spare (supernumerary) embryos as an appropriate source for stem cell research, in an interesting conditional formulation it deemed 'the creation of embryos with gametes donated for the purpose of stem cell procurement [as] ethically unacceptable, when spare embryos represent a ready alternative' (EGE, 2000: para 2.7; Figure 14.2, cells 5–8 plus a conditional acceptance of cells 1–4). Meanwhile, 'the creation of embryos by somatic cell nuclear transfer for research on stem cell therapy would be *premature*' since there are alternative sources (EGE, 2000: para 2.7, stress added; Figure 14.2, cells 13–16). Embedded in this discourse are notions, firstly, of embryo status contingent upon source and, secondly, of ethics as a developmental process that moves from 'premature' to, presumably, mature.

The EGE report signalled a move by some actors involved in the construction of the political narrative to attempt to change the debate from one characterised by static and opposing ethical positions to one where successive refinements of position were normal and negotiation possible. As time drew nearer for Parliament to consider the Commission's Framework Programme Six proposal, and as the critics of human embryo research made it clear that they would use this as an opportunity for expressing their opposition, so the objective need for negotiating room increased. Although the subsequent Parliamentary debate on the First Reading of the proposal in November 2001 suggests that little had changed, and that categorical statements of broad ethical positions were still the norm, the amendments incorporated into the proposal proved otherwise. The amendments meant that FP6 would not fund 'research activity aiming at human cloning for reproductive purposes' or 'the creation of embryos for research purposes including somatic cell nuclear transfer' (therapeutic cloning – Figure 14.2, cells 13–16). However, it would fund (and here is the compromise) 'research on "supernumerary" early-stage (i.e. up to 14 days) human embryos (embryos genuinely created for the treatment of infertility so as to increase the success rate of IVF but no longer needed for that purpose and when destined for destruction)' (European Parliament, 2001a: Article 3; Figure 14.2, cells 5–8).

The success of these amendments indicates that there is in the situation a sub-text of covert political negotiation around the ethical components of embryo source and embryo age (up to 14 days). (The latter is also described as the 'pre-embryo', an important political category in the long running UK embryo debate.) Under pressure from the conflicting political constituencies of the Framework Six Programme, the transnational narrative was beginning to evolve and to suggest that some types of embryos are ethically more important than others. In an attempt to facilitate this evolution and as part of the search for a way through the thickets of the ethical debate, in December 2001 the European Parliament set up the Temporary Committee on Human Genetics and Other New Technologies of Modern Medicine to report on the ethical, legal, economic and social implications of human genetics. In the event, its activities served to stimulate the involvement of new civil society policy networks in the discussion and legitimise the inclusion of fresh ethical dimensions. As the debate on its final report on 29 November 2001 demonstrates, ethical collisions in the Parliamentary arena were at this stage more achievable than were compromise positions (European Parliament, 2001b).

In contrast to this, in the separate arena of the EGE the expert agenda of human embryo research was experiencing a further process of ethical refinement in response to scientific and industrial demands for greater regulatory protection of their ESC investments. Reflecting on the issue of patenting, the Group stated its opinion that 'patenting of inventions allowing the transformation of unmodified stem cells from human embryonic origin into genetically modified stem cell lines or specific differentiated stem cell lines for specific therapeutic or other uses, is ethically acceptable as long as the inventions fulfil the criteria of patentability' (EGE, 2002: para 2.5). However, this liberalisation of the ethics of patenting was balanced by the EGE's view on therapeutic cloning where, drawing on its earlier Opinion *Ethical aspects of human stem cell research and use*, it called for 'a cautious approach, excluding the patentability of the process of creation of a human embryo by cloning for stem cells' (EGE, 2002: para 2.5).

It is clear that at this stage the search for practical ethical solutions to cultural conflict around ESCs was progressing much more swiftly in the expert arena of the bioethicists than in the Parliamentary and Council arenas of the politicians. Nonetheless, it was in the latter two arenas that a way forward had to be found if FP6 was to be funded. Institutional struggle was about to begin in earnest. In an interesting and, in the view of the opponents of human embryo research, challenging manoeuvre, at the Second Reading of the FP6 proposal in June 2002 Parliament voted through the overarching Framework Programme and transferred the issue of the criteria for embryo and human ESC research to the process for approval of the relevant Specific Research Programme (European Parliament, 2002). This meant that the Parliament was not directly involved in the decision-making because under EU procedures the Specific Programme details are a 'technical issue' and can be decided upon by Council without the agreement of Parliament. However, the advantage gained by this institutional move appeared to have been shortlived when in the September of the same year, under pressure from Austria, Italy, Germany and Ireland, the Council decided on a package of measures in response to the opposition concerns. This reiterated the ban on therapeutic cloning research and, furthermore, stipulated that there should be: a moratorium on the EU funding of human embryo and human ESC research until December 2003; a report on human embryonic stem cell research as the basis for an inter-institutional seminar on bioethics; and, taking into account the seminar's outcome, further guidelines on the principles that should guide Community funding of such research to be produced by December 2003 (European Parliament, 2002).

The increasing salience of novel modes of organised ethical engagement (Temporary Committee on Human Genetics, EGE Opinions, inter-institutional seminar, development of ethical funding guidelines, ethical review of projects) are indicators of the intensifying search for practical mechanisms for including cultural factors in the EU's transnational governance of the life sciences. However, the widespread recognition that cultural values are a legitimate component of the transnational governance of the life sciences did not readily lead to a parallel acceptance of the new mechanisms for the resolution of cultural conflict. Following the Commission's exhaustive report on human embryonic stem cell research and the inter-institutional seminar drawing on its findings in April 2003, the terms and constituency of the debate were undoubtedly enhanced – but so also was the difficulty of finding a sustainable compromise position. Despite further intense manoeuvrings, no progress was made and by default, therefore, the criteria contained in the European Parliament and Council Decision of 27 June 2002 and the Council Decision of 30 September 2002 in respect of the Specific Research Programme remained in place. Human embryonic stem cell research using therapeutic cloning (Figure 14.2, cells 13–16) could not be funded, that based on supernumerary embryos (Figure 14.2, cells 5–8) could, and the position of research using donated and aborted embryos (Figure 14.2, cells 1–4 and 9–12) as the source remained unresolved. FP6 could proceed, but so would the debate over the cultural values that should guide its human embryo research.

Conclusions

The cultural politics of FP6 and human ESC science are characterised by a continuing volatility in both the balance between the ethical components of the policy discourse and its possible policy outcomes. As the units of cultural trading on human embryo and ESC line research have become ever more refined in terms of source and creation date, so the range of possible compromises have multiplied. Yet at the same time as increasing the chances of political compromise, this situation has also increased the instability of any compromised achieved. Furthermore, there is no single policy discourse reflective of a particular set of cultural values capable of dominating the debate for sustained periods. Apparent triumphs by political actors in one institutional arena of the EU is swiftly countered by an outflanking move by their opponents in another arena. Alliances are unstable and porous.

Some parts of the political gameplay are nonetheless more volatile than others. Whereas the public debates of the European Parliament on human ESCs and FP6 were usually characterised by the stark presentation of conflicting cultural positions, in the expert arena of bioethics the search for compromise ethical equations has generated a quite different political style characterised by reason, flexibility and adaptation. While in the former, the cultural politics were raw and challenging, in the latter the explicit search for political utility has necessitated the development of the rules and procedures that can contribute to a practical outcome. Cultural politics in the EU is therefore operating at two levels in order to accommodate the otherwise incompatible requirements of (a) the unchanging legitimacy of particular value positions and (b) the need for those positions to be negotiable. As the application of the ethical components of Figure 14.2 to the political narrative of both levels has illustrated, there now exists a range of finely graded value positions on human ESC research that constitute the currency for political trading. Although at the public level this trade would be denied, the evidence of the political narrative is that such trading indeed occurs, though inefficiently, and will continue to occur as Framework Programme Seven stimulates similar concerns.

15
Regulation and the Positioning of Complementary and Alternative Medicine

John Chatwin and Philip Tovey

Introduction

Most developed and prosperous countries have seen their reliance on conventional medicine paralleled by an exponential growth of interest in complementary and alternative medicine (CAM) (Eastwood, 2000; McGregor and Peay, 1996). In this chapter, we discuss some of the results relating to research conducted by the authors that explored the ways in which CAM is used, understood and evaluated in terms of its perceived utility by patients with cancer in the UK and Pakistan. Our focus here, compared with earlier chapters, is less on formal processes of regulation and evaluation of IHTs, and more on the ways in which patients (and patient groups) position CAM in relation to orthodox medicine. We show how there are quite diverse ways in which this can occur that reflect different interpretive and interactional repertoires used by patients and CAM groups. Focusing on our UK data, we argue that the degree to which it is possible to build a genuinely 'alternative' medicine is constrained by organisational and gate-keeping constraints imposed on CAM groups.

In the UK and US, as well as in many other countries, there have been definite cultural moves towards embracing (or 'rediscovering') therapeutic traditions that engender models of health and illness that are often very different from those proposed by the dominant medical paradigm (see, for example, Eisenberg et al., 1993; Reilly, 1999; Sharma, 1992). There is still, however, widespread resistance within orthodox medicine to therapies that do not conform to current scientific models (see, for example, Fitzgerald, 1983; Lerner, 1984), nor meet the usual criteria of evaluation deployed in fields such as health technology assessment (Angell et al., 1998) even if it is also the case that clinical trials on CAM have been conducted (such as in the various US studies funded by the National Institute of Health: Sparber et al., 2000). The much quoted House of Lords Science and Technology Committee report on complementary and alternative medicine (House of Lords, 2000) categorised CAM modalities in terms of three groups: Group 1 included 'principle disciplines' such as acupuncture, homoeopathy, and herbal medicine. Group 2 included practices which did not purport to involve diagnostic skill, such as the Alexander Technique, aromatherapy, counselling, and stress therapy. Group 3 covered other alternative disciplines which lack any credible evidence base such as crystal therapy, iridology, radionics, dowsing and kinesiology.

Those (often holistic) systems that are tolerated by orthodoxy are rarely, if ever, fully assimilated on their own terms, and the tendency is for isolated elements or procedures to be 'cherry picked' depending on how well they can be shoe-horned into the dominant allopathic model. Aspects of alternative practice that have been incorporated have generally been appropriated with little reference to the knowledge base that produced them. Certain acupuncture techniques, for example, are now sometimes used in chronic pain management by orthodox surgeons and dentists, (Clinical Standards Advisory Group, 1999), but within conventional medicine there is little acknowledgement of the principles upon which acupuncture is based. Incorporation of the system has depended to some extent on the degree to which its fundamentally esoteric elements – such as the network of invisible meridian lines that form the basis of the acupuncturist's view of the human body – have been successfully explained away. The deeper, and many would argue, equally significant philosophical and holistic components that remain unexplainable in these terms are largely dismissed. It is interesting to note that in the UK, training in traditional acupuncture takes a number of years, and entry into an appropriate professional organisation (The Acupuncture Society, for example) requires a period of apprenticed training as well as proof of accreditation from an appropriate college, etc. However, anyone from a conventional medical background can take a short course in 'medical' acupuncture – an activity which, although it utilises needles and related acupuncture techniques, rejects fundamental aspects of the traditional knowledge system. The message this sends is that very little value is placed on the validity of the principles on which traditional acupuncture is based. This selectivity extends to other aspects of CAM too. Perhaps one of the biggest issues dogging the wider integration of therapies is the lack of acceptable (i.e. scientific) evidence relating to how they actually work. Proponents of CAM, however, can point to established orthodox treatments and procedures which, while they may be effective, are just as poorly understood.

CAM and integration

This is not to say that there have been no benefits – from the perspective of CAM activists at least – in terms of the raised profile that certain therapies have gained, and positive effects for patients accruing from the selective adoption of CAM by conventional medicine. There is now, for example, a growing movement within the medical profession that aims to 'imbue orthodox medicine with the values of complementary medicine' (Rees and Weil, 2001). The call for *integrated medicine* (or *integrative medicine* as it is in the US), although still grounded in an orthodox paradigm, has at least allowed those medical practitioners who might wish to explore the possibility of other approaches to do so more openly. Along with the direct adoption of 'holistic' principles, there have also been attempts by a significant number of orthodox practitioners to develop styles of consultation behaviour that, although they may not be directly acknowledged as such, have striking similarities to many of the principles underpinning overtly complementary approaches. The concept of *patient-centred* medicine has resonated with many doctors as being crucial to the delivery of high quality care (Mead and Bower, 2000). Boyd and Heritage (2005) point out that there is a growing literature aimed at teaching new doctors to conduct sensitive and complete medical interviews that encourage patients to 'reveal their observations, concerns, and fears' (2), although this advice is often given with the proviso that the activity should not take up too much of the doctor's time (Coulehan and Block, 1987). Similarly, in the late 1990s the *British Medical Journal* ran a series of articles focusing on the concept of *narrative based medicine* (see: Greenhalgh and Hurwitz, 1999; Hudson-Jones, 1999; Elwyn and Gwyn, 1999; Greenhalgh, 1999; Launer, 1999). This is an approach to consulting that aims to integrate more than the purely symptomatic information that a patient brings; the 'story' of the patient's illness, and how the illness fits into their life-world paradigm is incorporated – something that resonates strongly with many CAM modalities.

At both a 'grass-roots' and policy level too, the idea of 'concordance' is widely regarded as being of benefit to both patients and doctors (see, for example: Lask, 2002; Dickinson et al., 1999). Concordance is positioned as being the opposite of 'compliance' and is a non-authoritarian and negotiated approach to treatment giving which is engendered by 'an agreement reached after discussion between a patient and healthcare professional that respects the beliefs and wishes of the patient in determining whether, when, and how medicines are to be taken' (Bryan, 2002: 425). In calling for the adoption of this paradigm by all practitioners, however, Bryan also points out that compliance (and by implication, the traditional notion that 'doctor knows best') is still apparently widespread within the medical profession. Indeed, Armstrong (2005) has argued that concordance is simply the 'acceptable face of compliance': as he says, 'the goals remain the same but the technique is more subtle as patients are recruited to direct themselves in medicine-taking' (26).

While there is clearly, therefore, a need to be cautious about the sense in which doctor–patient relations have changed, what these general trends illustrate is that, regardless of whether or not orthodox medicine embraces the theoretical (i.e. holistic, and often spiritual) underpinnings of CAM, there is a growing movement that regards elements within the methods that complementary therapists *in general* utilise to have a therapeutic value. Exactly what these elements are and how they actually work is open to debate, but it appears that they are often largely connected (either directly or tangentially) to issues of interaction: to elements of equality, respect and empowerment as they relate to the relationship between practitioner and patient. Acknowledgement and development of these basic interactional elements does not necessarily demand a rejection of orthodox methods, or even the complete validation of CAM treatments, so it could be argued that they provide an ideal arena by which integration (or at least mutual understanding) can be encouraged.

Critics and supporters of alternative medicine have argued that a high proportion of the therapeutic effect that accrues from complementary methods may be generated by the process that permeates the 'therapeutic encounter' (see: Reilly (unpublished); Reilly, 2001), and that it is often qualities within the patient / practitioner interaction that somehow stimulate a naturally occurring healing response. It is significant, for example, that people who have experienced both holistic and orthodox approaches often draw a contrast between the different interactional environments that conventional and complementary medicine generate (see: Montbriand, 1998). Similarly, studies have found that many people are attracted to alternative therapies out of a desire for a more 'humanistic' approach (see: Furnham, 1996; Astin, 1998) and for those who become regular adherents of CAM, much of the appeal can be grounded in the perception that the meeting with their practitioner will embody interactional elements that have become attenuated in conventional medical encounters (Chatwin and Collins, 2002). On a very broad level, this can be reflected in the feeling that a complementary practitioner has, for example, more time to listen to what they have to say, or is somehow more able to be empathetic than their allopathic counterpart.

This level of generalisation, however, can be problematic – especially when the vast array of different therapies, processes, remedies and treatments that get categorised under the umbrella of 'CAM' is considered. Although modalities which most readily evoke the CAM ethos – such as homoeopathy – do have an overt focus on the quality of the therapeutic encounter, and are fundamentally holistic, there are any number of other CAMs which could equally be described as reductionist in their approach to the patient (in terms of the way in which little or no holistic information is required in order to diagnose and treat complaints).

In developing countries CAM (or 'traditional medicine' – TM) has not been routinely appropriated from outside (as is the case with much of the 'exotic' CAM popular in the West), apart from one or two notable exceptions, such as homeopathy which is one of the most widely used forms of

CAM in Pakistan and a relatively modern system of medicine originating in Germany. For the most part, however, CAM is part of an indigenous folk tradition, and the utilisation of therapies by people proceeds in a somewhat more 'market led' way. In Pakistan, for example, the process of accessing CAM / TM carries little of the pseudo-spiritual and 'lifestyle' connotations which are found in the West: a person who is ill and may wish to use, say, a *Hakim* (traditional healer) will simply shop around for one with a good personal reputation and treat the encounter as they would any other service that they have to pay for. They may also simultaneously consult a western doctor, or buy medicines from a pharmacy. Although this kind of selective utilisation is obviously also found in the West, in Pakistan the divide between CAM / TM and orthodox medicine is far more blurred, and has different dynamics. The choices people make over which one to use are very much more pragmatic – often relating to issues of cost (CAM / TM usually being a cheaper option), and which treatments are perceived to be effective for a given complaint, rather than to issues of holism or patient participation in the treatment process.

The apparent move towards integrative practice, then, opens up new and interesting issues about how integration is being operationalised: who has power to shape the process, what elements of a given approach are being prioritised, and how issues such as evidence are being used in this regard? It can be seen as a process marked by appropriation (as in the case of acupuncture, for example) and selective incorporation (as in the case of patient 'empowerment'). It also resonates, however, with the continuing marginalisation of those practitioners and elements of CAM modalities which are deemed inappropriate by orthodox standards.

Cancer care and support groups

Although the ongoing CAM / orthodox interchange is being played out across all areas of health care, there are certain conditions and illnesses which have tended to be particularly dynamic sites of engagement. Cancer is one such illness. At present it is reportedly at the forefront of moves to integrate CAM into mainstream medicine, with many patients utilising some form of CAM as part of their treatment (Ernst, 2000). Significantly, however, in a field of health sociology which has mainly been concerned with the individual and individualised engagement, it has proven to be one of the few arenas where these exchanges can be subordinate to wider group-based processes; in cancer care, patients routinely learn about and access CAM through their involvement in group activities (i.e. support-groups, informal networks and so on). This makes the development and dynamics of these organisations particularly pertinent, as in terms of integration they have a unique position in relation to both the individual consumers (i.e. group members), and mainstream health care systems, and can be seen to actively reflect the underlying processes of appropriation, marginalisation and selective incorporation that are being played out in the wider CAM / orthodox arena.

There are a wide variety of self-help groups, networks and charities concerned providing CAM services for cancer patients. These range from organisations which are essentially divisions of conventional health care (such as groups based in oncology units or hospitals), through to 'grass roots' groups that are independent and have little or no affiliation to local health networks. The authors' research in this area has categorised the plethora of group formats into two main types relative to mainstream health care structures:

> *Type 1 groups* are groups that were established organisations before they began providing CAM services.
> *Type 2 groups* are groups set up with an overtly holistic agenda.

For organisations of the first type, in which there are well established working processes (often formulated along 'traditional' socio-medical lines) the incorporation of selected CAM therapies does not have any significant impact on organisational direction or ethos. These groups are, in the UK, routinely affiliated to NHS hospitals or hospices, and are, therefore, likely to have a strong institutional grounding (they are frequently led or organised by medically trained individuals such as Macmillan nurses, and tend to embody high levels of medicalisation; they often occupy buildings that are directly attached to hospitals and evoke a vaguely institutional atmosphere – even if they are overtly trying to avoid this. The therapies on offer will almost always be of the kind that conventional medicine considers 'safe' (i.e. harmless). Similarly, CAM therapists who work with these groups will be subject to intense professional scrutiny, and will be subordinate to medical personnel; only those therapists with appropriate credentials, and those willing to play along with this system of accreditation, will be allowed near a patient. In these environments CAM takes on a secondary role; both in relation to orthodox medicine in general, and also in relation to the group. It becomes just one of a number of discrete activities that group members can choose to be involved in. Therapies are not incorporated as part of a wider holistic agenda, and in this way the utilisation of CAM proceeds very much in terms of the orthodox organisational framework (and by extension, the orthodox medical perspective). While individual therapies and therapists may well benefit from the exposure that involvement brings (particularly in terms of the legitimacy conferred by mainstream sanction), the kind of rich CAM / life-world narrative that can be observed when exploration of therapies proceeds in *ad hoc* and holistic environments is largely absent.

Type 2 groups are ostensibly different. These organisations are ones which have an overtly CAM or holistic agenda. Although generally much smaller, less well connected (in terms of wider health networks) and under-funded when compared to their NHS affiliated partners, their independence allows them to more readily reflect an underlying CAM ethos: CAM forms an integral part of what they do and they are generally run and organised by CAM therapists, or individuals with a commitment to particular therapies or holistic approaches. This holistic grounding (and the generally freer attitude towards group structures and hierarchies that it can engender), to some extent, frees them up from the restrictions of mainstream bureaucracy and officialdom, although even in the most extremely 'alternative' organisations there is generally an acknowledgement of the need to be selective about the extent to which conventional methods of group management are rejected. Type 2 groups are also apparently freer to be more accommodating towards the qualifications and training of the therapists that they utilise; they can offer CAM modalities that may be marginal and not represented by conventionally accredited training. Similarly, Type 2 groups generally engender a defined and separate 'CAM' identity which Type 1 groups do not, and which can provide a degree of cohesion and unity between members.

Given the prominent position that cancer care has as a harbinger of CAM integration, this apparently polarised group categorisation might appear to illustrate an open and evolving CAM integration process. What the dynamics and makeup of these two types of support group actually tell us about an underlying process of selective incorporation, however, is significant. While on the surface, it might appear that the very existence of groups in the Type 2 category indicate a liberalisation of orthodox attitudes and a wide choice of therapeutic modalities, the reality is somewhat different. Although these groups are free to offer any number of esoteric, marginal and (in the House of Lords terms) 'Group 3', practices, what actually tends to happen is that organisations which aspire to any degree of long term existence usually end up offering the same small portfolio of 'safe' therapies that are found in Type 1 groups (i.e. ones which are essentially toothless in terms of 'threat' value to orthodox medicine). These routinely include healing / therapeutic touch, aromatherapy, and physical activities such as yoga or tai chi, which are especially popular as they can be done in sessions involving a whole group simultaneously. It appears that organisations engage in a kind of self-generated selective incorporation. Even the most esoteric group needs to establish at least some links with mainstream health care systems in order to function. While it is the case that more radical groups find ways round the problem of remaining true to their ideals while at the same time continuing to access mainstream networks, without connections to the environments where patients are likely to be found – oncology clinics, for example – a group is doomed, or is at least forced to remain relatively insular; access implies legitimation, and access can be denied unless you play the game.

Similarly, there are issues of safety and efficacy; for a group that wants to attract and retain members, it becomes counter-productive to project an image which is too dismissive of orthodox positions, or too overtly 'alternative'. Many potential members will, after all, have little background knowledge of CAM, and only a minority are likely to embrace radical or 'rejectionist' perspectives – particularly at the onset of their engagement with CAM. So while CAM activists and adherents are on the face of it free to incorporate anything they choose into their therapeutic agendas (and thus help to project an image of cancer care as a beacon of integration) what actually happens is that their efforts to do this are subtly undermined by the tangential practicalities of continuing as organisations. Orthodox medicine remains firmly in control of the arena and while CAM integration (implying a degree of equality) is overtly welcomed, it is far from the coming together of equal partners that the orthodox likes to project – and the more naïve integrationists like to believe.

What we can draw from these examples then, is a picture of the relationship between CAM and orthodox modalities which is more complex than might at first appear, and one in which the response of conventional medicine to the threat posed by CAM is ever more subtle and sophisticated. In the face of a movement that can go well beyond the purely medical to incorporate elements of spirituality, lifestyle choice and socio-political awareness, the simplistic and ineffective response of outright rejection initially adopted has now been superseded by varying degrees of appropriation and incorporation. As always, however, nothing that threatens the underlying power base of conventional medicine will ever really be sanctioned, and 'doctor will always know best'.

Conclusion

In this chapter we have briefly discussed some of the dynamics that shape CAM in its relation to orthodoxy, and have suggested how this relationship informally acts to regulate and evaluate the position and thereby, in effect, *meaning and experience* of CAM for the patient and indeed practitioner (whether allopathic or alternative per se). While the interactional and holistic aspects of CAM have both a rhetorical and actual role to play in marking out the perceived and real characteristics of CAM compared with orthodox medicine, in practice these play a more secondary role in regard to the way CAM is evaluated, this being much more dependent on organisational and institutional constraints in the health care system.

16
Evaluation as an Innovative Health Technology
David Armstrong

Introduction

Explanations for the 'success' of new medical technologies have stressed the importance of stabilising social, political and natural readings and orderings of the world (Brown and Webster, 2004). Technologies do not succeed because they are inherently better at tackling a problem, rather their progress is the outcome of a number of factors that may or may not produce success. Central to this analysis of contingency is the idea of the network. New technologies are hybrids of human actors, natural phenomena and sociotechnical production that sustain or undermine their claims of superiority. An increasing number of studies have revealed the importance – and inherent uncertainties – of this networking process.

One consequence of the network approach to analysing the development and emergence of new technologies is that studies in the area tend towards individual case histories. Certainly, the generalisability of networks as a construct is reaffirmed but given the uncertainties surrounding the emergence of any specific new technology it is only by teasing out the idiosyncratic circumstances of its emergence that its unique course can be plotted. This produces detailed historical analyses of the success of new technologies but does little to identify more generalised approaches to technology development as any new analysis is condemned to trace the detailed steps of its subject. Yet as the emergence of new technologies is rooted in social processes it might seem reasonable to look for regularities at certain points in the network. In other words, it is possible that certain elements of networks become so formalised and stabilised that they establish a less contingent more routinised element to technology development. It is the exploration of such a network node that forms the basis of this paper.

Innovative health technologies

Until recently innovative health technologies (IHTs) were probably no different from any other technologies. A new diagnostic machine or therapeutic tool was embedded in networks that succeeded or failed to recruit support for the transition into everyday clinical practice. Le Fanu (1999) gives some sense of the trials and tribulations surrounding new medical technologies when he described the considerable costs for patients who, in effect, were the experimental test-bed on which the technologies such as renal dialysis were developed. In the mid 1970s, however, this approach to IHTs began to change with the emergence of evidence-based medicine (EBM) which promoted a formalised testing process for assessing whether new technologies actually worked (Sackett and Rosenberg, 1995; Sackett, Rosenberg, Gray et al., 1996).

The key claim of EBM was the existence of a hierarchy of evidence supporting the effectiveness of any technology. This hierarchy ranged from low-level descriptive studies to the randomised controlled clinical trial at the pinnacle. The essence of the randomised clinical trial was the use of a research design that attempted to eliminate bias from technology evaluations. Thus, by randomising a patient population into two groups, one of which received the new intervention and the other that did not, any differences in outcome could be confidently ascribed to the effect of the intervention rather than extraneous factors such as the placebo response, case mix of patients, etc.

The success of EBM in promoting trial methodology/technology can be gauged by widespread commitment to randomised clinical trial evaluation by regulatory bodies, professional groups and health care funders. In the UK this involves the presentation of trial evidence on effectiveness by pharmaceutical companies seeking regulatory approval; it involves the use of trial-based evidence for approving the use of new technologies in the NHS (through the National Institute of Clinical Excellence); it involves the Department of Health funding collation of trial evidence through the Cochrane Centre (since grown into an international Cochrane Collaboration); and it involves substantial investment in funding new trials such as through the Department of Health's Health Technology Assessment Programme, the Medical Research Council (MRC), the pharmaceutical industry and medical research charities. The triumph of EBM is to persuade these various interests that only the randomised trial can adequately test the effectiveness of any new IHT.

Evaluation techniques

The core of the new evaluation technology is both a hierarchy of knowledge reflecting the 'strength' of the evidence for the effectiveness of the technology and a formalised assessment pathway. IHTs first need to pass through the 'proof of concept' gateway, a preliminary demonstration that there is reasonable probability that they might be effective in patient populations. Then the IHT proceeds through Phase I and Phase II trials to establish safety, potential 'doses' and likely effectiveness before being tested in the Phase III randomised clinical trials. Phase III trials (or randomised clinical trials) were first used in a significant way in 1948; their value was promoted in Cochrane's celebrated 1972 monograph, *Effectiveness and Efficiency in the NHS*; they then achieved extensive 'success' with the emergence and growth of the EBM movement from the mid 1970s which recruited the randomised clinical trial as the centre-piece of evidence production.

Trial methodology has continued developing. The proclaimed goal has been the further elimination of bias so that trial methodology can better claim to produce a 'fair assessment' of any IHT. More sophisticated trial designs have been developed (factorial, cluster, preference, pragmatic, etc.), more ingenious statistical techniques applied, particularly to estimating the number of patients needed in any trial to produce a clear result (the so-called statistical power of a study), and improved practical measures to minimise bias, such as better blinding and third party randomisation, have been actively promoted. But there was one largely non-technical problem that trials have needed to engage with, namely the appropriate *outcome measure* to compare the relative success of intervention and control groups.

Until about 30 years ago health care interventions were evaluated according to 'biomedical' criteria. The most salient was mortality – did the intervention decrease the number of deaths amongst patients with the specified condition? – though other biomedical and physiological measures such as reduction in blood pressure in hypertension or increase in haemoglobin in anaemia were used to indicate that treatments 'worked'. But then, in the mid 1970s a new outcome measure emerged that apparently celebrated the patient's perspective over the doctor's. This new indicator of medical success was called 'quality of life' (hereafter, QoL).

QoL emerged at the conjunction of several social processes. Although QoL provided a new outcome measure for clinical trials it was not a product of such a direct need. Its origins lie in part in the pressures to find an outcome measure that addressed the patient's perspective. Biomedical outcomes seemed technical and often far removed from the patient's interests in having subjectively good health. The advent of an increasingly consumerist approach to health care delivery, especially in the traditionally paternalistic NHS, offered the political leverage to a search for a more patient-based assessment of health status.

The idea of QoL, however, emerged in a socio-political rhetoric that appeared after WWII to address a number of problems seemingly left in the wake of technological success. These problems can be grouped into three main strands: dilemmas of social progress; the price of technological success; and chronic illness and the quality of care (Armstrong and Caldwell, 2004).

During the second half of the twentieth century social scientists began to explore the use of 'social indicators' (measuring educational attainment, housing completions, health care use, crime, etc.) as a means of measuring social change. In part these indicators were intended to measure social progress but they also addressed the spectre of social crisis reflected in the reported increase in crime, drug use and social fragmentation which were occurring despite significant economic and social progress. The underlying issue was frequently identified as the two-edged sword of 'technology' which, on the one hand, was responsible for considerable social benefit and on the other for many social ills. As Watts observed, technology produced both 'the quality of life which is now becoming possible, but also ... the population growth and environmental pollution which make this quality of life and living quite impossible' (Watts, 1970). QoL therefore became the mediator between these two conflicting strands of the effect of new technologies: whether the benefits or dis-benefits were dominating could be summarised using the yardstick of the society's QoL.

A more specific example of the potential costs associated with apparently beneficial technologies was a series of wide-ranging developments in new medical technologies during the 1950s and 60s including renal dialysis, renal transplantation, new forms of cancer therapy and intensive care for neonates. While these IHTs meant that acute and life-threatening diseases and disorders could be addressed therapeutically (if not in terms of a complete 'cure') instances were increasingly recorded of 'successful' technological interventions that, overall, had a deleterious effect on patient well-being whether it was the pain and suffering of early transplant patients, the social and physical burden of the renal dialysis programme or the disabled children produced by neonatal interventions. In these cases technological success seemed to be bought at the expense of (chronic) 'human cost', a burden that fell not only on the individual patient but also on the health service and a wide range of informal carers. Again, as with the general societal effects of technological advance, the notion of QoL appeared as the crucial factor that would determine the success and limits of technological intervention. Such a balance of gains and losses was also to underpin the nascent medical ethics in its analysis of the appropriateness of medical interventions.

Finally, and relatedly, the growth in the numbers of patients burdened with chronic illness placed new pressures on the health and social care sectors. Patients who might have died in the past now survived but often with serious disabilities and those with serious disabilities had lengthening life spans. In addition, the numbers of elderly in the population were increasing, as a result both of medical intervention but also of fertility patterns earlier in the century. Crude measures of medical success such as survival were of only limited value in these circumstances. So, QoL seemed the ideal construct by which to assess the value of health care and the health status of the patient especially in those parts of health care, such as nursing, that relied on more intangible goals than the traditional biological ones of the medical profession (Najman and Levine, 1981). QoL as an outcome of care also provided a benchmark that could inform wider economic and political questions about the distribution of resources.

In summary, the origins of the QoL instruments were grounded in three quite distinct concerns relating to the paradoxes and contradictions of social progress, most notably the positive and negative effects of medical technologies on the very meaning of health beyond that defined in biomedical terms. What exactly constituted QoL was never clearly articulated; it was at once a construct embedded in everyday meanings and also a necessary concept that enabled some key problems of the post-war socio-political agenda to be addressed, at least in terms of rhetorical debate.

Towards a subjective measure of health

The development of empirical measures of QoL was achieved by combining various existing instruments and technologies (Armstrong, Lilford, Ogden and Wessely, submitted). The whole process depended crucially on questionnaire technology that had been developed earlier in the twentieth century as a means of accessing subjective responses (particularly in inter-war opinion polls). Questionnaires were deployed during WWII to make rapid and efficient psychiatric assessments, especially among soldiers. Post-war surveys of the mental health of whole communities confirmed the value of the psychiatric questionnaire. This early success at measuring a population's mental health was based on the way in which the psychiatric instrument could attempt to reproduce the psychiatric interview for, unlike the identification of organic disease, psychiatric diagnosis rarely required a clinical examination. And as 'psychological symptoms' were 'subjective' they could later be construed as measures of subjective health status; then, when subjective health was being merged with the concept of QoL, they could also become in the 1980s measures of QoL itself.

In psychiatric disease the symptom often was the 'pathology' (an anxious patient had 'anxiety', a depressed patient 'depression', etc). For non-psychiatric disease, however, symptoms were only rough indicators of what the underlying pathology might be; it was usually the clinical examination or investigations that enabled the definitive diagnosis. Therefore, while it would have been possible to construct a checklist of physical symptoms for patients to complete, the particular configuration of noted symptoms could not easily translate into a diagnosis as it did in the psychiatric field. But symptoms could be used prognostically, that is for assessing the impact and severity of the disease.

The symptom of pain had an important diagnostic function but in the 1960s it also took on an important prognostic role. If the symptom got worse in chronic disease, that is pain increased, then clearly the disease was also getting worse irrespective of what the 'objective' pathological tests might show. This shift from signs and laboratory investigations to reports of symptoms as guides to the progress of disease represented a new emphasis on the patient's perspective. Its measurement did not require a new 'scope' for peering inside the body but a technique that would allow the capturing and ordering of experience. The pain questionnaire was the next logical step.

If the symptom of pain could be measured by questionnaire then why not other symptoms? Symptoms that were distressing for patients might indicate the course of the underlying illness. Symptom checklists therefore began to emerge both as components of more ambitious instruments but also in their own right as ways of capturing the patient's QoL. Several patients might have rheumatoid arthritis, for example, but the common diagnostic label concealed a great diversity of patient incapacity. One means of assessing this range of disease impact was to devise new measures of the effect of the illness on patients' lives. In particular, the effect of disease on physical functioning became the new marker of illness severity. These became known generically as measures of activities of daily living (ADLs): basically they listed everyday activities and the patient or a carer or a health professional would score the patient's ability to carry out each named activity.

With the development of psychiatric instruments, symptom questionnaires and ADLs it was a relatively simple task to merge all three into a comprehensive questionnaire. At first these patient self-completion questionnaires were held to elicit 'subjective health' but very quickly, and particularly in retrospect, they became proto-measures of QoL.

It was a relatively small step to add a fourth 'dimension' to the trilogy of mental and physical functioning and specific symptom experiences. It was apparent that decline in both mental and physical functioning owed their impact to the effect on social functioning. It was therefore not a major change to add a few questions more directly on social functioning itself. The classic QoL instrument was now complete, and between about 1980 and 1995 gradually consolidated its hold over medical outcomes research and practice.

Between the mid 1970s when rather crude attempts were made to operationalise QoL in terms of whether the patient was mobile or could be returned home to the sophisticated multi-dimensional scales of the 1990s, the patient's health-related QoL has become a major indicator of the patient's health status. At the beginning of the twenty-first century there were over 800 different instruments in existence and their number continues to grow. But it was the recruitment of QoL measures to the new EBM movement that established its dominance as both an outcome measure and goal of health care. In short, these and similar initiatives in Europe and elsewhere, reflect the fact that QoL measurement has grown to become, arguably, the most important indicator of medical outcome in the opening years of the twenty-first century. The concept of QoL has a central place in formal health care evaluations.

Technology, networks and nodes

The above outline provides a schematic case history of the development of an evaluative IHT that incorporates the randomised clinical trial and QoL outcome measures. This outline can be seen to involve a network of different elements that came together to establish the dominant evaluation regime of the early twenty-first century. These elements included the consumerist movement that asserted the rights of patients to be heard and for their agenda to be the focus of health care delivery; they included the pre-existing technologies of questionnaire format and design; they included the incorporation and cannibalisation of pre-existing health status instruments; and they included the socio-political rhetoric that invented and placed such a premium of QoL in the second half of the twentieth century. At the conjunction of these various network elements the QoL instrument emerged. Whether it would have been other than a measurement curiosity is unclear without its linkage to the emerging evaluation technologies that were actively promoting the clinical trial as the gold-standard of assessment procedures. Thus, married to an increasingly dominant evaluation technology, the QoL instrument emerged as the pre-eminent measure of health status in the late twentieth and early twenty-first centuries and, in a reflexive gesture, the new goal of clinical practice.

Health care evaluation technology, however, is not static but continues to advance its main claim of providing fair assessments by devising ever more elaborate ways of eliminating bias. Yet perhaps more important, are the outcome measures this new technology incorporates which centre around QoL. For all its apparent simplicity and elegance QoL measurement remains – and will always remain – a contested arena (Siegrist and Junge, 1989; Fitzpatrick, 1996). QoL depends on social evaluations of the 'good life' and these will both change over time and are likely to differ for different people and groups in the population. In part this has been accommodated by the development of individualised measures in which the patient determines the quality criteria that will be used though these tend to be cumbersome to use and are not suited to forming the outcome measures of clinical trials (Hickey et al., 1996). But in more traditional instruments, whose social values are encoded (perhaps ill patients, healthy people, representative samples, minority groups? and so on), there remains a constant difficulty as new measures were devised to accommodate different social/illness groups. Then there are problems with determining which of the values that are elicited should be represented in the instrument, how they cluster together and how they should be asked. In summary, the conceptual and technical problems in measuring QoL that have been tackled over the last four decades provides a rich social history in its own right; but when this search for the perfect instrument is coupled with the greater role of QoL measures in determining successful IHTs then the conceptual and technical debates about QoL measurement achieve a much wider social significance.

While the success of QoL instruments can be described in terms of network processes, it is clear that QoL has itself become an important component of a wider IHT network. This new evaluative technology, supported by developments in evaluation methodology, provides the mechanism that enables other innovations to transfer from the laboratory to the clinic. Whereas in the past IHT networks behaved like any other, there is now an important 'constant' in their stabilisation mechanisms and, given the dominance of the evaluation technologies embodied in EBM and the randomised clinical trial, an essential gateway to clinical success. No doubt the role of networks is crucial in bringing IHTs to the point of clinical application but then they face the task of demonstrating their effectiveness through formalised evaluation whatever the strength of the network support behind it.

The new health care evaluation mechanisms such as EBM, randomised clinical trials and QoL measures can therefore be seen as forms of technology in their own right and also essential components of the network through which the success of other IHTs can be established. The EBM movement of the 1970s and its subsequent success can be viewed like any other IHT, as subject to the same forces and contingencies as any other technology. But the success of evaluation methodology/technology had increasingly profound implications for other IHTs. Certainly other IHTs influence each other in that they were all network components but the RCT was unique in claiming to constitute an autonomous network node through which all other technologies had to pass if their effectiveness was to be definitively established. In other words, whereas previous networks functioned as iterative inter-related elements that, in certain circumstances, produced an effective technology, EBM removed the uncertainty and contingency inherent in this process. EBM offered a technique for determining effectiveness relatively independently of other network elements.

Whereas according to actor-network theory the fact that a technology 'worked' was a product of whether it 'held'(a situation determined by success at recruiting and stabilising elements of a supportive network) (Latour, 1987), the claim of EBM is that effectiveness can be 'out-sourced', so to speak, to a different technological process. Thus the randomised clinical trial rather than the network can be used to demonstrate that a technology works.

Finally, because all new IHTs now include a formal evaluative component (since this is now a precondition for 'success'), often with QoL as an outcome, it can be seen that all new IHTs also embody that same evaluation technology. The success of the drug, the machine, the diagnostic test depends on successful integration with the evaluation technology. Much work in the social studies of science has emphasised the importance of the process of recruiting support to the successful 'stabilisation' of a new technology such that it will be accepted and used (Latour, 1987). In an important sense this network of support – people, resources, machines, objects, inscriptions (Akrich, 1992) – should be viewed as a part of the technology in question as without it the technology would not 'work'. Given the critical evaluative processes that IHTs must now undergo it would therefore seem reasonable to regard evaluative technology as an integral component of the underlying instrumental biomedical technology itself. Thus a heart transplant is not only the product of specialised surgical and immunological techniques but also of the evaluative processes that confirmed that it worked to the patient's advantage; the use of a new drug not only reflects on the pharmacological resources and skills that developed it in the laboratory but also the clinical trials that brought it through into everyday practice; a new body scanner is not only a machine based on fundamental physical principles but also an evaluative technology that showed that the machine could pass the test of being 'effective'.

Whether evaluative technologies are viewed as parallel to conventional IHTs or an integral part of them, they represent an important interface between IHTs and the social domain in as much as they increasingly determine the criteria by which any new technology will be judged.

Evaluative technologies – that increasingly emphasised the measurement of QoL – therefore ensured the integration of an essentially social dimension with the instrumentality of the health innovation. Biomedical technologies emerged from the basic 'natural' medical sciences such as biochemistry, pharmacology, engineering, genetics, etc. In marked contrast, the evaluative technologies are derived from social and population sciences, from statistics, economics, sociology, psychology and epidemiology. This means that the successful introduction and adoption of a new medical technology requires a fusion of the technological output of natural and social sciences, or at least for the output from the social sciences to act as a kind of legitimation for the everyday application of the IHT.

The way that the social sciences have approached the task of identifying an endpoint of health care that would best reflect the benefit of intervention to the patient as person, rather than the patient as simply anatomical body, was to 'encode' the person's view into a formal QoL instrument. In encoding subjective views of what constitutes a good quality life, QoL measures have brought a notional patient's perspective to the success or failure of the instrumental IHT. An IHT may 'work' in the laboratory but in failing the QoL test would be 'shown' not to work in the world outside.

Conclusion

Evaluation technology in medicine now relies on the primacy of the randomised clinical trial and to an increasing extent on QoL as the arbiter of 'success'. This technology is subject to the same forces as other IHTs in its growth and development and therefore can be considered as yet another IHT. And yet this particular technology has claims to control the technology development process in ways that other IHTs cannot match. Thus, despite its own instabilities, evaluation technology has successfully colonised a part of the network in which other IHTs are invested. In effect, part of the developmental network has become formalised as an essential passage through which other IHTs now must pass before they can be considered for generalised clinical application.

The project on which this chapter is based has mapped the emergence of QoL over the last few decades as a major new outcome for clinical practice and research. In part it is an illustration of how IHTs emerge. In part, however, this research also provides an account of a revolutionary new medical technology that is not only instrumentally applied by medicine but comes to change the very nature of clinical practice and control the destinies of other far removed IHTs.

Bibliography

Aanestad, M., Rotnes, J. S., Edwin, B., and Buanes, T. (2002) From operating theatre to operating studio – visualizing surgery in the age of telemedicine. *Journal of Telemedicine and Telecare* 8: 56–60.

Abraham, J. (1995) *Science, Politics and the Pharmaceutical Industry: Controversy and Bias in Drug Regulation*. London, UCL/St. Martin's Press.

Abraham, J. (2002a) The pharmaceutical industry as political player. *The Lancet* 360: 1498–502.

Abraham, J. (2002b) Making regulation responsive to commercial interests: streamlining drug industry watchdogs. *British Medical Journal* 325: 1164–69.

Abraham, J. (2003) The science and politics of medicines control. *Drug Safety* 26: 135–43.

Abraham, J. (2004) Pharmaceuticals, the state and the global harmonisation process. *Australian Health Review* 28: 150–61.

Abraham, J. and Davis, C. (2005) A comparative analysis of drug safety withdrawals in the UK and the US (1971–1992): implications for current regulatory thinking and policy. *Social Science & Medicine* 61: 881–92.

Abraham, J. and Lewis, G. (2000) *Regulating Medicines in Europe: Competition, Expertise and Public Health*. London, Routledge.

Abraham, J. and Reed, T. (2001) Trading risks for markets: the international harmonisation of pharmaceuticals regulation. *Health, Risk & Society* 3: 113–28.

Abraham, J. and Reed, T. (2002) Progress, innovation and regulatory science: the politics of international standard-setting. *Social Studies of Science* 32: 337–69.

Abraham, J. and Reed, T. (2003) Reshaping the carcinogenic risk assessment of medicines. *Social Science & Medicine* 57: 195–204.

Abraham, S. and Llewellyn-Jones, D. (1997) *Eating Disorders: The Facts*. Oxford, Oxford University Press.

Abramsky, L., Hall, S., Levitan, J., and Marteau, T. M. (2001) What parents are told after pre-natal diagnosis of a sex chromosome abnormality: interview and questionnaire study. *British Medical Journal* 322: 463–66.

Abramson, A. (1950) The veteran paraplegic problem as I see it. *Paraplegia News* April: 5.

Adam, B. (1994/1990) *Time and Social Theory*. Cambridge, Polity Press.

Adam, B. (1995) *Timewatch: The Social Analysis of Time*. Cambridge, Polity Press.

Adkins, L. (2002), *Revisions: Gender and Sexuality in Late Modernity*. Buckingham, Open University Press.

Akrich, M. (1992) The de-scription of technical objects, in W. Bijker and J. Law (eds), *Shaping Technology/Building Society*. Cambridge, MIT Press.

Alfirevic, Z. and Neilson, J. P. (2004) Antenatal screening for Down's syndrome. *British Medical Journal* 329: 811–12.

Allen, K. and Williamson, R. (2000) Screening for hereditary haemochromatosis – should be implemented now. *British Medical Journal* 320: 183.

Anderson, J. Innovation and locality: hip replacement in Manchester and the northwest. *Bulletin of the John Rylands University Library*, forthcoming 2006.

Anderson, J. and Timmermann, C. (eds) (2006) *Designs and Devices: Medical Technologies in Historical Perspective*, Palgrave, forthcoming 2006.

Anderson, T. (1956) Paraplegia in review, open letter to Dr. Henry L. Heyl. *Paraplegia News* December: 8.
Andersson, F. (1992) The drug lag issue: the debate seen from an international perspective. *International Journal of Health Services* 22: 62–8.
Angell, M. and Kassirer, J. (1998) Alternative medicine – the risks of untested and unregulated remedies. *New England Journal of Medicine* 339: 839–41.
Annandale, E. and Clark, J. (1996) What is gender? Feminist theory and the sociology of human reproduction. *Sociology of Health and Illness* 18, 1: 17–44.
Anon (1987) Pan-European regulatory body inevitable, says ABPI. *Scrip* 1213: 1–3.
Anon (1988a) UK Meds Division – proposed changes. *Scrip* 1279: 3.
Anon (1988b) Swedes want SLA changes. *Scrip* 1285: 6.
Anon (1988c) New license fees proposed. *Scrip* 1369: 8.
Ansell Pearson, K. (1999). *Germinal Life*. London, Routledge.
Apple, A. J. and Sims, J. (1996) Harold Ridley and the invention of the intraocular lens. *Surveys of Ophthalmology* 40: 279–92.
Apple, D. J. and Mamalis, M. (1984) Complications of intraocular lenses: a historical and histopathalogical review. *Surveys of Ophthalmology* 29: 1–54.
Armstrong, D. (1995) The rise of surveillance medicine. *Sociology of Health and Illness* 17: 393–404.
Armstrong, D. (2002) *A New History of Identity*. Basingstoke, Palgrave Macmillan.
Armstrong, D. (2005) The myth of concordance: response to Stevenson and Scambler. *Health* 9: 23–7.
Armstrong, D. and Caldwell, D. (2004) Origins of the concept of quality of life in health care: a rhetorical solution to a political problem. *Social Theory and Health* 2: 361–71.
Armstrong, D., Lilford, R., Ogden, J. and Wessely, S. Health-related quality of life and the transformation of symptoms. Submitted.
Ashworth, M., Clement, S., and Wright, M. (2002) Demand, appropriateness and prescribing of lifestyle drugs: a consultation survey in general practice. *Family Practice* 19(3).
Astin, J. A. (1998) Why patients use alternative medicine: results of a national study. *Journal of the American Medical Association* 279: 1548–53.
Audit Commission (2000) *Fully Equipped: the provision of equipment to older or disabled people by the NHS and social service in England and Wales*. London.
Ball, M. J. and Lillis, J. (2001). E-health: transforming the physician/patient relationship. *International Journal of Medical Informatics* 61: 1–10.
Barlow, J., Bayer, S. and Curry, R. (2004) Organisational complexity or technological complexity? Innovation Studies Centre, Tanaka Business School, Imperial College London.
Bashshur, R. L. (1997) Critical Issues in Telemedicine. *Telemedicine Journal* 3: 113–26.
Barad, K. (1998) Getting real: technoscientific practices and the materialization of reality. *Differences: A Journal of Feminist Cultural Studies* 10: 88–128.
Barnartt S. and Scotch, R. (2001) *Disability Protests: Contentious Politics, 1970–1999*. Washington DC, Gallaudet University Press.
Barnes, C. (1991) *Disabled People in Britain and Discrimination*. London, Hurst & Co.
Barraquer, J. (1956) The use of plastic lenses in the anterior chamber – indications-techniques-personal results. *Transactions of Ophthalmological Society of the UK* 74: 537–52.
Barraquer, J. (1959) Anterior chamber plastic lenses. results of and conclusions from five year's experience. *Transactions of the Ophthalmological Society of the UK* 79: 393–424.

Barton Hutt, P. (1995) A history of government regulation of adulteration and misbranding of medical devices. *Food, Drug, Cosmetic Law Journal* 44: 99–117.
Barry, A. (2001) *Political Machines: Governing a Technological Society*. London, The Athlone Press.
Bayer, S., Barlow, J., and Curry, R. (2004) Assessing the impact of a care innovation: telecare [paper] Tanaka Business School, Imperial College London.
Beck, U. (1992) *Risk Society: Towards a New Modernity*. London, Sage.
Beck, U. (1998) Politics of risk society, *in* Franklin, J. (ed.), *The Politics of Risk Society*. Cambridge, Polity.
Beck, U. and Beck-Gernsheim, E. (2002), *Individualisation: Institutionalised Individualism and its Social and Political Consequences*. London, Sage.
Beck-Gernsheim, E. (2002) Health and responsibility: from social change to technological change and vice versa, *in* B. Adam, U. Beck and J. Van Loon (eds) *The Risk Society and Beyond: Critical Issues for Social Theory*. London, Sage.
Berg, M. and Mol, A. (1998) (eds) *Differences in Medicine: Unravelling Practices, Techniques and Bodies*. Durham NC, Duke University Press.
Bessell, T. L., Silagy, C. A., Anderson, J. N., Hiller, J. E., and Sansom, L. N. (2002) Quality of global E-Pharmacies: can we safeguard consumers? *European Journal of Clinical Pharmacology* 58: 567–72.
Beutler, E., Felitti, V. J., Koziol, J. A., Ho, N. J. and Gelbart, T. (2002) Penetrance of 845G-A (C282Y) HFE hereditary haemochromatosis mutation in the USA. *The Lancet* 359: 211–18.
Bharadwaj, A. (2002) Uncertain risk: genetic screening for susceptibility to haemochromatosis. *Health, Risk and Society* 4: 227–40.
Bijker, W. E. (1995) *Of Bicycles, Bakelite and Bulbs: Toward a Theory of Sociotechnical Change*. Cambridge, MIT Press.
Billings, C. F. (1998) Attendant and observer in the early days of the physically disabled students' program and the Center for Independent Living, 1969–1977, *in* K. Cowan (ed.) *Builders and Sustainers of the Independent Living Movement in Berkeley*, vol. 2, The Bancroft Library, University of California, Berkeley, 2000.
Binkhorst, C. D. (1959) Iris-supported artificial pseudophakia. A new development in intra-ocular artificial lens surgery. *Transactions of the Ophthalmological Society of the UK* 79: 569–84.
Bloor, D. (1978) Polyhedra and the abominations, Leviticus. *British Journal for the History of Science* 39: 245–72.
Blume, S. S. (1992) *Insight and Industry: on the Dynamics of Technological Change in Medicine*. Cambridge, MIT Press.
Blume, S. (1995) *Cochlear Implantation: Establishing a Clinical Feasibility, 1957–1982 in* N. Rosenberg, et al. (eds) *Sources of Medical Technology*. Washington, National Academy Press.
Blumenthal, D. (2002) Doctors in a wired world: can professionalism survive connectivity? *The Milbank Quarterly* 80: 525–46.
Bognor, W. C. (1996) *Drugs to Market*. London, Pergamon Press.
Booth, B. and Zemmel, R. (2004) Prospects for productivity. *NRDD* 3: 451–6.
Bordo, S. (1993) *Unbearable Weight: Feminism Western Culture and the Body*. Berkeley, University of California Press.
Bothwell, T. H. and Macphail, A. P. (1998) Hereditary haemochromatosis: etiologic, pathologic and clinical aspects. *Seminars in Hematology* 35, 55–71.
Boukes, F. S., Groeneveld, F. P. M. J. et al. (2003) Dutch College of general practitioners' position on hormone replacement therapy (HRT). http://ntg.artsennet.nl/sitemap 2005 (22 February).

Bourke, J. (1996) *Dismembering the Male: Men's Bodies, Britain and the Great War*, London, Reaktion Books.
Boyd, E. A. and Heritage, J. (2005) Analyzing history-taking in primary care: questioning and answering during the verbal examination. In J. Heritage and D. Maynard (eds), *Practicing Medicine: Talk and Action in Primary-care Encounters*. Cambridge, Cambridge University Press.
Bracken, P. and Thomas, P. (2001) Postpsychiatry: a new direction for mental health. *British Medical Journal* 322: 724–7.
Brett, A. S. (1984) Ethical issues in risk factor intervention. *American Journal of Medicine* 76: 557–61.
Briggs, E. (2001) Interview with Brian Woods, Waterlooville, 5 September 2001.
British Medical Association (2001) *Withholding and Withdrawing Life – Prolonging Medical Treatment*. London, BMA.
British Medical Journal (2003) Public retains great trust in doctors, MORI poll shows http://www.bma.org.uk/ap.nsf/Content/MORI03.
Broom, A. F. (2004) The E-male: prostate cancer masculinity and online support as a challenge to medical expertise. *Journal of Sociology* 41:1.
Brown, N. and Michael, M. (2004) Risky creatures: institutional species boundary change in biotechnology regulation. *Health, Risk and Society* 6: 207–22.
Brown, N., Rappert, B. et al. (eds) (2000) *Contested Futures: a Sociology of Prospective Techno-Science*. Aldershot, Ashgate.
Brown, N. and Webster, A. (2004) *New Medical Technologies and Society: Reordering Life*. Cambridge, Polity Press.
Bruch, H. (1973) *Eating Disorders: Obesity, Anorexia and the Person Within*. New York, Basic Books.
Brumberg, J. J. (2000) *Fasting Girls: The History of Anorexia Nervosa*. New York, Vintage.
Bryan, L. (2002) Daily regimen and compliance with treatment: concordance respects beliefs and wishes of patients. *BMJ* 324: 425.
Bryant L., Green J. M. and Hewison, J. (2001) Prenatal screening for Down's syndrome: some implications for a 'screening for all' policy. *Public Health* 115: 356–8.
Bryant, L. D, Green, J. M. and Hewison, J. (2005) Attitudes towards prenatal diagnosis and termination in women who have a sibling with Down's syndrome. *Journal of Reproductive and Infant Psychology* 23: 179–96.
Bryant, L., Green, J. M., Hewison, J., Sehmi, I. and Ellis, A. (2001) Descriptive information about Down's syndrome: a content analysis of serum screening leaflets. *Prenatal Diagnosis* 21: 1057–63.
Bud, R. (1995) In the engine of industry: regulators of biotechnology, 1970–86, *in* M. Bauer (ed.). *Resistance to New Technology: Nuclear Power, Information Technology and Biotechnology*. Cambridge, Cambridge University Press, 293–310.
Burke, W., Thomson, E., and Khoury, M. J. (1998) Hereditary haemochromatosis: gene discovery and its implications for population-based screening. *JAMA* 280: 172–8.
Burrows, R. (2000) Virtual community care? Social policy and the emergence of computer mediated social support. *Information, Communication & Society* 3: 95–121.
Bury, M. (1998) Postmodernity and health, in G. Scambler and Paul Higgs (eds) *Modernity, Medicine and Health*. London, Routledge.
Byng, S. and Hewitt, A. (2003) From doing to being: from participation to engagement, *in* S. Parr, J. Duchan, and C. Pound, *Aphasia Inside Out*. Maidenhead, Open University Press.
Cahill, S. E., and Eggleston, R. (1994) Managing emotions in public: the case of wheelchair users. *Social Psychology Quarterly* 57: 300–12.

Callon, M., Law, J., and Rip, A. (eds) (1986) *Mapping the Dynamics of Science and Technology*. London, Macmillan – now Palgrave Macmillan.
Campbell J. and Oliver, M. (1996) *Disability Politics: Understanding Our Past Changing Our Future*, London, Routledge.
Cane, J. (1971) Minute sheet correspondence from J. Cane to D. A. R. Peel, 5 August 1971: File MT97/1168, Public Record Office, Ruskin Avenue, Kew, Richmond, TW9 4DU, UK.
Carter, S. (1995) Boundaries of danger and uncertainty: an analysis of the technological culture of risk assessment, in J. Gabe (ed.) *Health, Medicine and Risk: the Need for a Sociological Approach*. Oxford, Blackwell, 13–50.
Centre for Medicines Research (CMR) International (2002) *CMR International News* 20(1).
Charatan, F. (2002) Buyer beware remains US policy towards information on the Net. *British Medical Journal* 324: 566.
Charles River Associates (2004) *Innovation in the Pharmaceutical Sector: a Study Undertaken for the European Commission*. London.
Chatwin, J. and Collins, S. (2002) Communication in the homoeopathic consultation. *The Homoeopath* 84: 24–7.
Chilaka, V. N., Konje, J. C., Stewart, C. R. Narayan, H. and Taylor, D. J. (2001) Knowledge of Down's syndrome in pregnant women from different ethnic groups. *Prenatal Diagnosis* 21: 159–64.
Chisholm, K. (2002) *Hungry Hell – What it's Really Like to be Anorectic: a Personal Story*. London, Short Books.
Choyce, D. P. (1960) The use of all-acrylic anterior chamber implants. *Transactions of the Ophthalmological Society of the UK* 80: 201–19.
Clark, E. A. (1995) Population screening for genetic susceptibility to disease. *British Medical Journal* 311: 35–8.
Clinical Standards Advisory Group (1999) *Services for Patients with Pain*. London: Department of Health.
Clough, K. and Jardine, I. (2003). Telemedicine: five years on – what progress? *The British Journal of Healthcare Computing & Information Management* 20: 21–3.
CMR International (2005) Innovation on the wane? *Latest News*. Available at http://www.cmr.org
Cochrane, A. L. (1972) *Effectiveness and Efficiency in the NHS*. Oxford, Nuffield.
Collins, H. M. and Evans, R. (2003) King Canute meets the beach boys: responses to the third wave. *Social Studies of Science* 33: 435–52.
Conrad, P. (1985) The meaning of medications: another look at compliance. *Social Science and Medicine* 20: 29–37.
Corbin, J. and Strauss, A. (1985) Managing chronic illness at home: three lines of work. *Qualitative Sociology* 8: 224–47.
Corby, P. (2002) Interview with Brian Woods, London, 9 May 2002.
Coulehan, J. L. and Block. M. (1987) *The Medical Interviewer: A Primer for Students of the Art*. Philadelphia, F. A. Davis Company.
CPMP. Committee for Proprietary Medicinal Products (2001) *Points to consider on the manufacture and quality control of human somatic cell therapy medicinal products*. CPMP/BWP/41450/98.
Craigie, M., Muncer, S., Loader, B. and Burrows, R. (2002) Reliability of health information on the Net: an examination of experts' ratings. *Journal Of Medical Internet Research* 4: 1
Cranney, A. (2003) Treatment of postmenopausal osteoporosis. *BMJ* 327: 355–6.

Crawford, R. (1980) Healthism and the medicalization of everyday life. *International Journal of Health Science* 19: 365–88.

Cullen, J. (1998) The needle and the damage done: research, action research and the organisational and social construction of health in the 'information society'. *Human Relations* 51: 12.

Cullen, J. (2004) The information brothel: collaborative learning and the collapse of professionalism *in*, E. Harlow and S. Webb (eds) *Information and Communication Technologies in the Welfare Services*. London: Jessica Kingsley.

Daly, K. J. (1996) *Families and Time: Keeping Pace in a Hurried Culture*. Thousand Oaks, Sage.

Davis, M., Hart, G., Imrie, J., Davidson, O., Williams, I. and Stephenson, J. (2002), 'HIV is HIV to me': meanings of treatments, viral load and re-infection among gay men with HIV. *Health, Risk and Society* 4: 31–43.

Davis-Floyd, R. W. (1994) The technocratic body: American childbirth as cultural expression. *Social Science and Medicine* 38: 1125–40.

Davison, C., Macintyre, S. and Davey-Smith, G. (1994) The potential social impact of predictive genetic screening for susceptibility to common chronic diseases: A review and proposed research agenda. *Sociology of Health and Illness* 16: 340–71.

Deleuze, G. and Guattari, F. (1988) *A Thousand Plateaus: Capitalism and Schizophrenia* London, Athlone Press.

Department of Health (1998a) *Genetics and Cancer Services: Report of a Working Group*. London, Department of Health.

Department of Health (1998b) *Information for Health: an Information Strategy for the Modern NHS 1998–2005*. London, Department of Health.

Department of Health (2001a) *A Code of Practice for Tissue Banks: Providing Tissues of Human Origin for Therapeutic Purposes*. London, Department of Health.

Department of Health (2001b) *The Expert Patient: a New Approach to Chronic Disease Management for the 21st Century*. London, Department of Health.

Department of Health (2002) *Delivering 21st Century IT Support for the NHS*. London, Department of Health.

Department of Health (2004) *NHS Improvement Plan 2004: Putting People at the Heart of Public Services*. London, Department of Health.

DG Enterprise (2004) *Proposal for a Harmonised Regulatory Framework On Human Tissue Engineered Products*. Brussels, DG Enterprise, European Commission.

Dias, K. (2003) The Ana Sanctuary: women's pro-anorexia narratives in cyberspace, accessed via: http://www.bridgew.edu/SoAS/jiws/April03/Dias FINAL.pdf

Dickinson, D., Wilkie, P. and Harris, M. (1999) Taking medicine: concordance is not compliance. *BMJ* 319: 787.

Dingwall, R. (1994) Litigation and the threat to medicine, *in* J. Gabe, D. Kelleher and G. Williams (eds) *Challenging Medicine*. London, Routledge.

Douglas, M. (1966) *Purity and Danger: An Analysis of Concepts of Pollution and Taboo*. London, Routledge and Kegan Paul.

Doward, J. and Reilly, T. (2003) How macabre world of the Web offers fresh insight on anorectics. *The Observer*, 17 August 2003. Accessed at: http://observer.guardian.co.uk/uk_news/story/0,6903,1020383,00.html

Drews, J. (2000) Drug discovery: A historical perspective. *Science* 287: 1960–4.

Eastwood, H. (2000) Why are Australian GP's using alternative medicine? Postmodernisation, consumerism and the shift towards holistic health. *Journal of Sociology* 36, 133–5.

Eder, K. (1996) *The Social Construction of Nature*. London, Sage.

Edwards, A., Elwyn, G., Atwell, C., Wood, F., Prior, L. and Houston, H. (2005) Doing shared decision making and risk communication in practice: qualitative study of general practitioners' experiences in an explanatory trial. *British Journal of General Practice* 55 (510): 6–13.

Eedy, D. J. and Wootton, R. (2001) Teledermatology: a review. *British Journal of Dermatology* 144: 696–707.

Eisenberg, D. M., Kessler, R. C., Cindy, F., Norlock, F., Calkins, D. R. and Delbanco, T. L. (1993) Unconventional medicine in the United States – prevalence, costs, and patterns of use. *The New England Journal of Medicine* 328: 246–52.

Ellershaw, J. and Ward, C. (2003) Care of the dying patient: the last hours and days of life. *BMJ* 326: 30–4.

Elman, R., Parr, S. and Moss, B. (2003) The Internet and aphasia: crossing the digital divide, *in* S. Parr, J. Duchan, and C. Pound (eds) *Aphasia Inside Out*. Maidenhead, Open University Press.

Elwyn, G. and Gwyn, R. (1999) Stories we hear and stories we tell: analysing talk in clinical practice. *BMJ* 318: 186–8.

EMEA (2004) *Ninth Annual Report 2003*. London: European Agency for the Evaluation of Medicinal Products.

Epstein, S. (ed.) (1996) *Impure Science: AIDS, Activism and the Politics of Knowledge*. Berkeley, University of California Press.

Epstein, S. (2000), Whose identities, which differences? Activism and the changing terrain of biomedicalisation. Paper presented at HIV and Related Diseases Conference (HARD) Social Research Conference, Sydney, 13 May.

Erni, J. (1992) Articulating the (im)possible: popular media and the cultural politics of 'curing AIDS'. *Communication* 13: 39–56.

Ernst, E. (2000) Unconventional cancer therapies. *Chest* 117: 307–08.

Ettorre, E. (2000) Reproductive genetics, gender and the body: 'Please doctor, may I have a normal baby?' *Sociology* 34: 403–20.

EU Commission (2003a) Directive 2003/63/EC of 25 June 2003 amending Directive 2001/83/EC of the European Parliament and of the Council on the Community code relating to medicinal products for human use. Official Journal L 159, 27/06/2003 P. 0046–0094.

European Commission (2003b) *Report on Human Embryonic Stem Cell Research*. SEC(2003)441, 3 April.

European Commission (2005) Special Eurobarometer. *Social Values, Science and Technology*. Available at: http://europa.eu.int/comm/public_opinion/archives/ebs/ebs_225_report_en.pdf

EU Commission DG Enterprise (2002) *Need for a Legislative Framework for Human Tissue Engineered Products*. http://europa.eu.int/comm/enterprise/medical_devices/consult_tissue_engineer.htm (accessed July 2002).

EU Commission DG Enterprise (2003) *Results of the Consultation on the Need for a Community Legal Framework on 'Human Tissue Engineered Products'* http://europa.eu.int/comm/enterprise/medical_devices/consultation_results.pdf. (accessed March 2003).

EU Commission DG Enterprise (2004a) *Proposal for a Harmonised Regulatory Framework on Human Tissue Engineered Products*: DG Enterprise Consultation Paper, 6 April 2004. Presented at Stakeholders' Conference, 16 April 2004, Brussels.

EU Commission DG Enterprise (2004b) *Proposal for a Harmonised Regulatory Framework on Human Tissue Engineered Products*. Summary of contributions. April 2004.

EU Commission DG Sanco (2003) *Amended proposal for a European and Council Directive on setting standards of quality and safety for the donation, procurement, testing, processing, storage, and distribution of human tissues and cells.* COM (2003) 340 final May 2003. http://europa.eu.int/eur-lex/en/com/pdf/2003/com2003_0340en01.pdf

Europabio, Eucomed, & Emerging Biopharmaceutical Enterprises (2004) *Joint industry comments on Proposal for a harmonised regulatory framework on human tissue engineered products.* DG Enterprise consultation paper 13 August 2004. http://www.europabio.be/positions/hTEPs0804.pdf (accessed 12/04).

European Council (1999). Council Decision 1999/167/EC Adopting a Specific Programme for Research, Technological Development and Demonstration on Quality of Life and Management of Living Resources, 1998–2002, *Official Journal.* L64, 12/3/99: 1–19.

European Group on Ethics in Science and New Technologies (2000) *Ethical Aspects of Human Stem Cell Research and Use.* Opinion no. 15 to the European Commission.

European Group on Ethics in Science and New Technologies (2002) *Ethical Aspects of Patenting Inventions Involving Human Stem Cells.* Opinion no. 16 to the European Commission.

European Parliament (2000) Resolution on human cloning, PE T5-0375/2000.

European Parliament (2001a), Proposal for a European Parliament and Council decision concerning the multiannual framework programme 2002–06 of the European Community for research, technological development and demonstration activities aimed at contributing towards the creation of the European Research Area. COM(2001)94, 14 November.

European Parliament (2001b) Debates of the European Parliament. Sitting of 29 November 2001. Human genetics.

European Parliament and the European Council (1999) Decision 182/1999/EC of the European Parliament and Council concerning the Fifth Framework Programme of the European Community for research, technological development and demonstration activities (1998–02). *Office Journal.* L26. 1/2/99.

Evans, J. (2002). *Playing God. Human Genetic Engineering and the Rationalisation of Public Bioethical Debate.* Chicago, University of Chicago Press.

Eysenbach, B. and Kohler, A. (2002) How do consumers search for and appraise health information on the world wide web? Qualitative study using focus groups, usability tests, and in-depth interviews. *British Medical Journal* 324: 573–7.

Faulkner, A., Geesink, I., Kent, J. and FitzPatrick, D. (2003) Human tissue engineered products – drugs or devices? *British Medical Journal* 326: 1159–60.

Faulkner, A., Kent, J., Geesink, I. and FitzPatrick, D. (2004) Purity and the dangers of innovative therapies: re-ordering regulation and governance in the shaping of tissue-engineered medical technology. Paper presented to EASST/4S Conference 'Public Proofs', École des Mines, Paris.

FDA (2004a) *Innovation or Stagnation: Challenge and Opportunity on the Critical Path to New Medical Products.* US Department of Health and Human Services.

FDA (2004b) http:// www.fda.gov/cder/rdmt/NMEapps93–04.htm

Featherstone, M. and Hepworth, M. (1991) The mask of ageing and the postmodern lifecourse, *in:* M. Featherstone, M. Hepworth and B. S. Turner (eds) *The Body: Social Process and Cultural Theory.* London, Sage.

Feenberg, A. (1991) *Critical Theory of Technology.* New York, Oxford University Press.

Feenberg, A. (1992) On being a human subject: interest and obligation in the experimental treatment of incurable disease. *The Philosophical Forum* 23 (3): 213–30.

Feenberg, A. (1996) *From Essentialism to Constructivism: Philosophy of Technology at the Crossroads.* San Diego, University of San Diego Press.

Finch, T., May, C., Mair, F., Mort, M., and Gask, L. (2003). Integrating service development with evaluation in telehealthcare: an ethnographic study. *British Medical Journal* 327: 1205–09.
Finkler, K. (2000) *Experiencing the New Genetics: Family and Kinship on the Medical Frontier.* Philadelphia, PA, University of Pennsylvania Press.
Finkler, K. (2001) The kin in the gene: the medicalization of family and kinship. *Current Anthropology* 42 (2): 235–63.
Fitzgerald, F. T. (1983) Science and scam: alternative thought patterns in alternative health care. *New England Journal of Medicine* 309: 1066–77.
Fitzpatrick, R. (1996) Alternative approaches to the assessment of health-related quality of life, in A. Offer (ed.) *In Pursuit of the Quality of Life.* London, Clarendon Press.
Fitzpatrick, R., Shortall, E., Sculpher, M., Murray, D., Morris, R. and Lodge, M. (1998) Primary total hip replacement surgery: a systematic review of outcomes and modelling of cost-effectiveness associated with different prostheses. *Health Technology Assessment* 2 (20).
Flower, R. (2004) Lifestyle drugs: pharmacology and the social agenda. *Trends in Pharmacological Sciences* 25: 182–5.
Flowers, P. (2001) Gay men and HIV/AIDS risk management. *Health* 5: 50–75.
Flowers, P., Davis, M., Hart, G., Imrie, J., Rosengarten, M. and Frankis, J. (2004) Diagnosis and identity amongst HIV positive Black Africans living in the UK (in press).
Food and Drug Administration (2001) *PHS Guideline on Infectious Disease Issues in Xenotransplantation,* 29 January.
Førde, O. H. (1998) Is imposing risk awareness cultural imperialism? *Social Science and Medicine* 47: 1155–9.
Foucault, M. (1975) *The Birth of The Clinic: an Archaeology of Medical Perception* New York, Vintage Books.
Foucault, M. (1977) The politics of health in the eighteenth century, in. *Power/Knowledge: Selected Interviews and Other Writings, 1972–1977,* C. Gordon (ed.), London, Harvester Press.
Foucault, M. (1979) *Discipline and Punish.* New York, Vintage Books.
Fox, N. J., Ward, K. J. and O'Rourke, A. J. (2005a) 'Expert patients', pharmaceuticals and the medical model of disease: the case of weight loss drugs and the internet. *Social Science and Medicine* 60 (6): 1299–309.
Fox, N. J., Ward, K. J. and O' Rourke, A. J. (2005b) Pro-anorexia, weight loss drugs and the Internet: an anti-recovery explanatory model of anorexia. *Sociology of Health and Illness* 27 (7), 944–71.
Fox, N. J, Ward, K. J. and O'Rourke, A. J (2006) A sociology of technology governance for the information age. The case of pharmaceuticals, consumer advertising and the Internet. Submitted to *Sociology* 40.
Fox, R. C. (2002) Medical uncertainty revisited. In *Gender, Health and Healing: The Public/Private divide.* G. Bendelow, M. Carpenter, C. Vautier and S. Williams. London, Routledge: 236–53.
Fox, R. C. and Swazey, J. P. (1978) *The Courage to Fail: a Social View of Organ Transplants and Dialysis.* Chicago, University of Chicago Press.
Fox, R. C. and Swazey, J. P. (1992) *Spare Parts: Organ Replacement in American Society.* New York, Oxford University Press.
Frank, A. (1995) *The Wounded Storyteller.* Chicago, University of Chicago Press.
Franklin, S. (2001) Culturing biology: cells lines for the new millennium. *Health* 5: 335–54.

Franklin, S. (2003) Rethinking nature-culture: anthropology and the new genetics. *Anthropological Theory* 3 (1): 65–85.
Franklin, S. and Lock, M. (eds) (2003) *Remaking Life and Death – Towards an Anthropology of the Biosciences*. Santa Fe, SAR Press.
Franklin, S. and Roberts, C. (2002) End of Award Report, 'Genetic information in the context of preimplantation genetic diagnosis', Economic and Social Research Council, Swindon.
Freeman, C. (1982) *The Economics of Industrial Innovation*. London, Pinter.
Freidson, E. (1970) *Profession of Medicine: a Study of the Sociology of Applied Knowledge*. New York, Harper Row.
Furnham, V. C. (1996) Why do patients turn to complementary medicine? An empirical study. *British Journal of Clinical Psychology* 35: 37–48.
Fuss, M. (1997) Attendant for Cowell Residents, Assistant Director of the Physically Disabled Students' Program, 1966–1972, in S. Bonney (ed.) Builders and Sustainers of the Independent Living Movement in Berkeley, Volume II, Regional Oral History Office, The Bancroft Library, University of California, Berkeley, 2000.
Gabe, J. (1995) Medicine, health and risk: the need for a sociological approach, in J. Gabe (ed.) *Health, Medicine and Risk: Sociological Approaches*. Oxford, Blackwell: 1–18.
Galambos, L. and Sturchio, J. L. (1998) Pharmaceutical firms and the transition to biotechnology: a study in strategic innovation. *Business History* 72: 250–78.
Gann, B. (1998). Empowering the patient and public through information technology, in J. Lenaghan (ed.) *Rethinking IT and Health*. London: Institute for Public Policy Research, 123–8.
Gannon, L. (1999). *Women and Aging: Transcending Myths*. New York, Routledge.
Gastaldo D. (1997) Is health education good for you? Re-thinking health education through the concept of bio-power, *in*, A. Petersen and R. Bunton (eds) *Foucault, Health and Medicine*. London, Routledge: 113–33.
Gelijns, A. C. and Rosenberg, N. (1995) The changing nature of technology development, *in* A. C. Gelijns and N. Rosenberg (eds), *Sources Of Medical Technology: Universities and Industry*. Institute of Medicine, Washington, National Academy Press.
Gelijns, A. and Rosenberg, N. (1999) *Diagnostic Devices: an Analysis of Comparative Advantages*, *in* D. Mowery and R. Nelson, (eds), *Sources of Industrial Leadership*. Cambridge, Cambridge University Press.
General Medical Council (2002) *Withdrawing and Withholding Life: Prolonging Treatments: Good practice in decision making. Guidance from the Standards Committee of the General Medical Council*. London, GMC.
Gibbons, M., Limoges, C. and Nowotny, H. (1994) *The New Production of Knowledge*. London, Sage.
Giddens, A. (1984) *The Constitution of Society*. Cambridge, Polity Press.
Giddens, A (1990) *The Consequences of Modernity*. Cambridge, Polity Press.
Giddens, A. (1994) *Modernity and Self-identity: Self and Society in the Late Modern Age*. Stanford, Stanford University Press.
Gifford, S. M. (1986) The meaning of lumps: A case study of the ambiguities of risk, *in* C. R. Janes, R. Stall and S. M. Gifford (eds) *Anthropology and Epidemiology*. Dordrecht, Reidel: 213–46.
Gilbert, D., Walley, T., and New, B. (2000) Life style drugs. *British Medical Journal* 321: 1341–4.
Gillett, J. (2003) Media activism and Internet use by people with HIV/AIDS. *Sociology of Health and Illness* 25: 608–24.

Ginsburg, F. and Rapp, R. (1995) *Conceiving the New World Order: the Global Politics of Reproduction*. Berkeley, CA, University of California Press.

Ginsburg, F. and Rapp, R. (2001) Enabling disability: rewriting kinship, re-imagining citizenship. *Public Culture* 13: 533–56.

Ginsburg, F. and Rapp, R. (2002) Facing disability as an un/imaginable cultural event. Paper presented at the American Anthropological Meetings, New Orleans, 21–24 November.

Glendinning, C., Kirk, S., Guiffrida, A. and Lawton, D. (2001) Technology-dependent children in the community: definitions, numbers and costs. *Child, Health and Development* 27: 321–34.

Goldsmith, S. (1963) *Designing for the Disabled*. London, Royal Institute of British Architects.

Goode, J., Greatbatch, D., O'Cathain, A., Luff, D., Hanlon, G., and Strangleman, T. (2004) Risk and the responsible health 'consumer': the problematics of entitlement among callers to NHS direct. *Critical Social Policy* 210–32.

Goode, J., Hanlon, G., Luff, D., O'Cathain, A., Strangleman, T., and Greatbatch, D. (2004a) Male callers To NHS Direct: the assertive carer, the new dad, and the reluctant patient. *Health* 311–28.

Goodman, J. (2000) Pharmaceutical industry, in J. V. Pickstone and R. J. Cooter, (eds) *Medicine in the 20th Century*. London, Harwood Academic Publishers.

Gordon, R. A. (2000) *Eating Disorders: Anatomy of a Social Epidemic*. Oxford, Blackwell.

Gott, M., Seymour, J. E., Bellamy, G., Clark, D. and Ahmedzai, S. (2004) How important is dying at home to the 'good death'? *Palliative Medicine* 18: 460–7.

Gottweis, H. (2002) Stem cell policies in the United States and Germany: between bioethics and regulation. *Policy Studies Journal* 30: 444–69.

Graber, M. A., Roller, C. M., and Kaeble, B. (1999) Readability levels of patient education material on the World Wide Web. *Journal Of Family Practice* 48: 58–61.

Grabowski, H., Vernon, J. and Thomas, L. (1978) Estimating the effects of regulation on innovation: an international comparative analysis of the pharmaceutical industry. *Journal of Law and Economics* 21: 133–63.

Gramsci, A. (1971) *Selections from the Prison Notebooks*. London, Lawrence and Wishart.

Greatbatch, D., Hanlon, G., Goode, J., Strangleman, T., Luff, D. and O'Cathain, A. (2006) Telephone triage, expert systems and clinical expertise. *Sociology of Health and Illness* 27 (6): 802–30.

Green, E. and Thompson, D. et al. (2002) Narratives of risk: women at midlife, medical 'experts' and health technologies. *Health Risk and Society* 4: 243–86.

Green, J. M., Hewison, J., Bekker, H. L., Bryant, L. D. and Cuckle, H. S. (2004) Psychosocial aspects of genetic screening of pregnant women and newborns: a systematic review. *Health Technology Assessment* 8: 33.

Green, R. M. (2001). *The Human Embryo Research Debates: Bioethics in the Vortex of Controversy*. Oxford, Oxford University Press.

Greenhalgh, T. (1999) Narrative based medicine in an evidence based world. *BMJ* 318: 323–25.

Greenhalgh, T, and Hurwitz. B. (1999) Why study narrative. *BMJ* 318: 48–50.

Gremillion, H. (2003) *Feeding Anorexia: Gender and Power at a Treatment Center*. North Carolina, Duke University Press.

Griffiths, F. E. and Green, E. et al. (2005) The nature of medical evidence and its inherent uncertainty for the clinical consultation: the example of midlife women. *BMJ* 311: 511.

Griffiths, F. E. and Jones, K. (1995) The use of hormone replacement therapy: results of a community survey. *Family Practice* 12: 163–5.

Griffiths, J. M., Bryar, R. M., Closs, S. J., Cooke, J., Hostick, T. and Kelly, S. et al. (2001) Barriers to research implementation by community nurses. *British Journal of Community Nursing* 6: 500–11.

Grimes, C. A. (1998) Attendant in the Cowell Residence Program, Wheelchair Technologist, and Participant/Observer of Berkeley's Disability Community, 1967–1990s, an oral history conducted in 1998 by David Landes, Regional Oral History Office, The Bancroft Library, University of California, Berkeley, 2000.

Grimes, C. A. (2002) Interview with Nick Watson, Berkeley, 9 October 2002.

Grogan, S. (1999) *Body Image*. London, Routledge.

Gutman, E. and Gutman, C. (1968) *Wheelchair to Independence: Architectural Barriers Eliminated*. Illinois, Springfield.

Habermas, J. (1984) *The Theory of Communicative Action*. London, Heinemann.

Habermas, J. (1990) *The Philosophical Discourse of Modernity*. Cambridge, MIT Press.

Hacking, I. (1990) *The Taming of Chance*. Cambridge, Cambridge University Press.

Hailey, D., Roine, R., and Ohinmaa, A. (2002) Systematic review of evidence for the benefits of telemedicine. *Journal of Telemedicine and Telecare* 8: 1–30.

Hallam, E., Hockey, J. and Howarth, G. (1999) *Beyond the Body: Death and Social Identity*. London, Routledge.

Hamilton, R. C. (2000) Sir Harold Ridley, MD, FRCS, FRS: inventor of the intraocular lens implant. *Current Anaesthesia and Critical Care* 11: 314–19.

Hancher, L. and Moran, M. (1989) Introduction, *in* L. Hancer and M. Moran (eds) *Capitalism, Culture and Economic Regulation*. Oxford, Clarendon.

Hanlon, G., O'Cathain, A., Goode, J., Luff, D., Strangleman, T. and Greatbatch, D. (2003) *NHS Direct and Patient Empowerment*. ESRC Final Report.

Hanlon, G., Strangleman, T., Goode, J., Luff, D., O'Cathain, A., and Greatbatch, D. (2005) Knowledge, technology and nursing: the case of NHS Direct. *Human Relations* 58: 147–71.

Hannay, D. R. (1979) *The Symptom Iceberg*. London, Routledge and Kegan Paul.

Haraway D. (1991a) *Simians, Cyborgs and Nature*. London, Free Association Books.

Haraway, D. (1991b) The Actors are Cyborg, Nature is Coyote, and the Geography is Elsewhere: Postscript to 'Cyborgs at Large', *in* C. Penley and A. Ross (eds) *Technoculture*. Minneapolis, University of Minnesota Press.

Haraway, D. (1997) *Modest_Witness@Second_Millenium.FemaleMan Meets_OncoMouse: Feminism and Technoscience*. London, Routledge.

Hardey, M. (1999) Doctor in the house: the Internet as a source of lay health knowledge and the challenge to expertise. *Sociology of Health & Illness* 21: 820–35.

Hardey, M. (2001) The story of my illness: personal accounts of illness on the Internet. *Health* 6: 31–46.

Hart, A., Henwood, F. and Wyatt, S. (2004) The role of the Internet in patient–practitioner relationships: findings from a qualitative research study. *Journal of Medical Internet Research* September, 6 (3): e 36. Available online: http://www.jmir.org/2004/3/e 36/

Haux, R., Ammenwerth, E., Herzog, W. and Knaup, P. (2002). Health care in the information society. A prognosis for the year 2013. *International Journal of Medical Informatics* 66: 3–21.

Heaton, J., Noyes, J., Sloper, P. and Shah, R. (2003a) Technology-dependent children and family life. *Research Works* 2003–02, York, Social Policy Research Unit, University of York.

Heaton, J., Noyes, J., Sloper, P. and Shah, R. (2003b) *Technology and Time: Home Care Regimes and Technology-Dependent Children*. York, Social Policy Research Unit, University of York.

Heaton, J., Noyes, J., Sloper, P. and Shah, R. (2005a) Families' experiences of caring for technology-dependent children: a temporal perspective. *Health and Social Care in the Community* 13 (5): 441–50.

Heaton, J., Noyes, J., Sloper, P. and Shah, R. (in press 2005b) The experiences of sleep disruption in families of technology-dependent children living at home. *Children and Society*.

Hedgecoe, A. and Martin, P. (2003) The drugs don't work: expectations and the shaping of pharmacogenetics. *Social Studies of Science* 33: 327–64.

Heidegger, M. (1977) *The Question Concerning Technology*. Trans. W. Lovitt. New York, Harper and Row.

Henderson, R., Orsenigo, L. and Pisano, G. P. (1999) The pharmaceutical industry and the revolution in molecular biology: interactions among scientific, institutional, and organizational change. In *Sources of Industrial Leadership*, ed. D. C. Mowery and R. Nelson. New York, Cambridge University Press, pp. 267–311.

Henwood, F., Wyatt, S., Hart, A. and Smith, J. (2003) Ignorance is bliss sometimes: constraints on the emergence of the 'informed patient' in the changing landscapes of health information. *Sociology of Health and Illness* 25: 589–607.

Heyman, B. and Henrikson, M. (2001) *Risk, Age and Pregnancy: a Case Study of Prenatal Genetic Screening and Testing*. Basingstoke, Palgrave – now Palgrave Macmillan.

Hickey, A. M., Bury, G., O' Boyle, C. A., Bradley, F., O'Kelly, F. D. and Shannon, W. (1996) A new short form individual quality of life measure (SEIQoL-DW): application in a cohort of individuals with HIV/AIDS. *British Medical Journal* 313: 29–33.

Hirji, J. (2004) Freedom or folly? Canadians and the consumption of online health information. *Information, Communication and Society* 7, 4 Special Issue on e-health.

Hn Tjura, A. (2000) A technological mediation of the medical-nursing boundary. *Sociology of Health and Illness* 22: 721–41.

Hoberman, M., Erbert, F., Cicenia, M. and Offner, E. (1952) Wheelchairs and wheelchair management. *American Journal of Physical Medicine* 32: 67–84.

Hobson, D. (2002) Reflections on rehabilitation engineering history: are there lessons to be learned? *Journal of Rehabilitation, Research and Development* 39, Supplement: 17–22.

Hochschild, A. (1983) *The Managed Heart: Commercialization of Human Feeling*. Berkeley, University of California Press.

Hockenberry, J. (2001) *Moving Violations: War Zones, Wheelchairs, and Declarations of Independence*. New York: Hyperion Books.

Hoffman, B. (2001) On the value-ladenness of technology in medicine. *Medicine, Health Care and Philosophy* 4: 335–46.

Hogle, L. F. (2002) Claims and disclaimers: whose expertise counts? *Medical Anthropology* 21: 275–306.

Holohan, A. (1977) Diagnosis: the end of transition, *in* A. Davis, and G. Horobin (eds) *Medical Encounters: Experience of Illness and Treatment*. London, Croom Helm.

Hood, C., Rothstein, H., and Baldwin, R. (2001) *The Government of Risk: Understanding Risk Regulation Regimes*. Oxford, Oxford University Press.

Horton, R. (2003) *Second Opinion: Doctors, Diseases and Decisions in Modern Medicine*. London, Granta Books.

House of Lords Science and Technology Select Committee (2000) *Sixth Report – Alternative Medicine*. London, House of Lords.

Hudson-Jones, A. (1999) Narrative in medical ethics. *BMJ* 318: 253–6.

Hughes, K., Bellis, M. and Tocque, K. (2002) *Public Health Information and Communication Technologies: Tackling Health and Digital Inequalities in the Information Age*. Liverpool, Centre for Public Health.

Hughes, T. P. (1983) *Networks of Power: Electrification in Western Society, 1880–1930.* Baltimore, Johns Hopkins University Press.
Hunt, D. L., Haynes, R. B., et al. (1998) Effects of computer-based clinical decision support systems on physician performance and patient outcomes: a systematic review. *JAMA* 280: 1339–46.
Illich, I. (1976) *Limits to Medicine. Medical Nemesis: the Expropriation of Health.* Harmondsworth, Penguin.
Illman, J. (2000) *The Expert Patient.* London: The Association of the British Pharmaceutical Industry.
Jacob, F. (1998) *Of Flies, Mice and Men.* Cambridge, Harvard University Press.
Jacobson, J. (1973) Letter to the editor. *Paraplegia News* November: 2.
Jaffe, N. S. (1999) Thirty years of intraocular lens implantation: the way it was and the way it is. *Journal of Cataract & Refractive Surgery*, Guest Editorial, 25 (4) April.
Jasanoff, S. (1995) Product, process, or programme? Three cultures and the regulation of biotechnology, in M. Bauer (ed.) *Resistance to New Technology.* Cambridge, Cambridge University Press.
Jasanoff, S., Markle, G. E., Petersen, J. C. and Pinch, T. (eds) (1995) *Handbook of Science and Technology Studies.* Thousand Oaks, CA, Sage.
Jaspers, K. (1963) *General Psychopathology* (7th edn) (trans. J. Hoenig and M. W. Hamilton). Manchester: Manchester University Press.
Jevons, F. (1992) Who wins from innovation? *Technology Analysis & Strategic Management* 4: 399–412.
Kagan, A. (1998) Supported conversation for adults with aphasia: methods and resources for training conversation partners. *Aphasiology* 12: 816–30.
Kaitin, K. I. and Di Masi, J. (2000) Measuring the pace of new drug development in the user fee era. *Drug Information Journal* 24: 673–80.
Kamenetz, H. (1969) A brief history of the wheelchair. *Journal of the History of Medicine and Allied Science* 24: 205–10.
Kanski, J. J. and Crick, M. P. (1977) Lensectomy *Transactions of the Ophthalmological Society of the UK* 97: 52–7.
Kaplan, B. (2001a) Evaluating informatics applications – clinical decision support systems literature review. *International Journal of Medical Informatics* 64: 15–37.
Kaplan, B. (2001b) Evaluating informatics applications – some alternative approaches: theory, social interactionism, and call for methodological pluralism. *International Journal of Medical Informatics* 64: 39–56.
Kaufman, H. E. (1980) The correction of aphakia. *American Journal of Ophthalmology* 89: 1–10.
Keith, R. D. F., Beckley, S., Garibaldi, J. M., Westgate, J., Ifeachor, E. C. and Greene, K. R. (1995) A multicentre comparative study of 17 experts and an intelligent computer system for managing labour using the cardiotogram. *British Journal of Obstetrics and Gynaecology* 102: 688–700.
Keith, R. D. F and Greene, K. R. (1994) Development, evaluation and validation of an intelligent system for the management of labour. *Bailliere's Clinical Obstetrics and Gynaecology* 8: 583–605.
Keith, R. D. F., Westgate, J., Hughes, G. W., Ifeachor, E. C. and Greene, K. R. (1994) Preliminary evaluation of an intelligent system for the management of labour. *Journal of Perinatal Medicine* 22: 345–50.
Keller, R. (2005) The origins of new drugs. *Nature Biotechnology* 21: 529–30.
Kelman, C. D. (1973) Phaco-emulsification and aspiration: a report of 500 cases. *American Journal of Ophthalmology* 75: 764–8.

Kelman, C. D. (1991) History of phacoemulsification. In M. D. Devine and W. Banko (eds), *Phacoemulsification Surgery*. London, Pergamon.
Kendall, L. (2001). *The Future Patient*. London, Institute of Public Policy Research.
Kennedy, J. and Saunders, D. (2002) *Mapping Survey of Down's Screening in London, Antenatal Screening Programme*. London Region.
Kent, J., Faulkner, A., Geesink, I. and FitzPatrick, D. (in press 2005) Towards governance of human tissue engineered technologies in Europe: framing the case for a new regulatory regime. *Technology Forecasting and Social Change*.
Kerr, A. (2004) *Genetics and Society: a Sociology of Disease*. London, Routledge.
Kerr, A. and Shakespeare, T. (2002) *Genetic Politics: From Eugenics to Genome*. Cheltenham, New Clarion Press.
Kessler, D. A., Hass, A. E., Feiden, K. L., Lumpkin, M. and Temple, R. (1996) Approval of new drugs in the US: comparison with the UK, Germany and Japan. *Journal of the American Medical Association* 276: 1826–31.
Khosa, J. (2003) Still life of a chameleon: aphasia and its impact on identity, *in* S. Parr, J. Duchan, and C. Pound, (eds) *Aphasia Inside Out*. Maidenhead, Open University Press.
Klecun-Dabrowska, E., and Cornford, T. (2000). Telehealth acquires meanings: information and communication technologies within health policy. *Information Systems Journal* 10: 41–63.
Klenerman, L. (ed.) (2002) *The Evolution of Orthopaedic Surgery*. London, Royal Society of Medicine Press Ltd.
Klinge, I. (1997) Brittle bones: medical interventions in osteoporosis, *in* K. Davis (ed.) *Embodied Practices: Feminist Perspectives on the Body*. London, Sage, pp. 59–72.
Kneller, R. (2005) National origins of new drugs. *Nature Biotechnology* 23: 655–6.
Knorr-Cetina, K. (1999) *Epistemic Cultures: How the Sciences Make Knowledge*. Cambridge, Harvard University Press.
Kong Y. Then (2003) The need to regulate human tissue engineered products: ophthalmology perspective. *British Medical Journal* 5 June 2003.
Konrad, M. (2005) *Narrating the New Predictive Genetics: Ethics, Ethnography and Science*. Cambridge, Cambridge University Press.
Lash, S. (2002) *Critique Of Information*. London, Sage.
Lask, B. (2002) Daily regimen and compliance with treatment: concordance respects beliefs and wishes of patients. *BMJ* 324–425.
Last, J. M. (1963) The iceberg 'completing the clinical picture' in general practice. *The Lancet* 282 (7297) 28–31.
Latour, B. (1987) *Science in Action*. Buckingham, Open University Press.
Latour, B. (1993) *We Have Never Been Modern*. Cambridge, Harvard University Press.
Latour, B. (2004) Why has critique run out of steam? From matters of fact to matters of concern. *Critical Inquiry* 30: 225–48.
Launer, J. (1999) A narrative approach to mental health in general practice. *BMJ* 318: 117–19.
Lauritzen, S. O. and Sachs, L. (2001) Normality, risk and the future: implicit communication of threat in health surveillance. *Sociology of Health and Illness* 23: 497–516.
Law, J. (1994) *Organising Modernity*. Oxford, Blackwell.
Lawrence, G. (ed.) (1994) *Technologies of Modern Medicine*. London, Science Museum.
Le Fanu, J. (1999) *The Rise and Fall of Modern Medicine*. London, Little, Brown and Co.
Lefebvre, R. C. Hursey, K. G. and Carleton, R. A. (1988) Labelling of participants in high blood pressure screening programs: implications for cholesterol screening. *Annals of Internal Medicine* 148: 1993–7.

Lehoux, P., Sicotte, C., Denis, J. L., Berg, M., and Lacroix, A. (2002) The theory of use behind telemedicine: how compatible with physicians' clinical routines? *Social Science and Medicine* 54: 889–904.
Lenaghan, J. (1998). *Rethinking IT and Health*. London, Institute for Public Policy Research.
Lerner, I. J. (1984) The whys of cancer. *Cancer* 53 (3) (sup): 815–19.
Lewando-Hundt, G., Shoham-Vardi, I., Beckerleg, S., Belmaker, I., Kassem, F. and Abu Jafar, A. (2001) Knowledge, action and resistance: prenatal screening amongst the Bedouin of the Negev, Israel. *Social Science and Medicine* 52(4): 561–9.
Lindsay, C. (2000) *Conquering Anorexia: the Route to Recovery*. Sussex, Summersdale.
Linebarger, E. L. and Hardten, D. R. et al. (1999) Phacoemulsification and modern cataract surgery. *Surveys of Ophthalmology* 44: 123–47.
Ling, R. (2002) *The development of the Exeter hip, in* J. Faux (ed.) *After Charnley*. Preston, The John Charnley Trust: 146–69.
Lippman, A., (1999) Choice as a risk to women's health. *Health, Risk and Society* 1: 281–91.
Loader, B. D., Muncer, S., Burrows, R., Pleace, N. and Nettleton, S. (2002) Medicine on the line? Computer-mediated social support and advice for people with diabetes. *International Journal Of Social Welfare* 11: 53–65.
Lock, M. (2001) On making up the good-as-dead in a utilitarian world, *in* S. Franklin and M. Lock (eds) *Remaking Life and Death*. New Mexico, School of American Research Press.
Lock, M. and Farquhar, J. (2005) *Beyond the Body Proper: Reading the Anthropology of Material Life*. Durham, Duke University Press.
Luck, J., Morgan J. F., Reid, F., O'Brien, A., Brunton, J. and Price, C. (2002) The Scoff questionnaire and clinical interview for eating disorders in general practice: a comparative study. *British Medical Journal* 325: 755–6.
Lupton, D. (1997) Foucault and the medicalisation critique, *in* A. Petersen and R. Bunton (eds) *Foucault, Health and Medicine*. London, Routledge: 94–110.
Lupton, D. (2002) Consumerism, reflexivity and the medical encounter, *in* M. MacSween, *Anorectic Bodies: a Feminist and Sociological Perspective on Anorexia Nervosa*. London, Routledge.
Lupton, D. and Seymour, W. (2000) Technology, selfhood and physical disability. *Social Science and Medicine* 50: 1851–62.
Lynn, J. and Adamson, D. M. (2003) *Living Well to the End of Life: Adapting Health Care to Serious Chronic Illness in Old Age*. Arlington VA, Rand Health.
MacKenzie, D. (1998) *Knowing Machines*. Cambridge, MIT Press.
MacKenzie, D. and Wajcman, J. (1999) *The Social Shaping of Technology*, 2nd edn Buckingham, Open University Press.
Macmillan Cancer Relief (2002) *Directory of Complementary Therapy Services in UK Cancer Care: Public and Voluntary Sectors*. London, Macmillan Cancer Relief.
MacSween, M. (1993) *Anorexic Bodies: a Feminist and Sociological Perspective on Anorexia Nervosa*. London, Routledge.
Majone, G. (1990) Introduction, *in* G. Majone (ed.) *Deregulation or re-regulation?* London, Pinter.
Majone, G. (1994) The rise of the regulatory state in Europe. *West European Politics* 17: 77–101.
Majone, G. (1996) *Regulating Europe*. London, Routledge.
Mair, F. and Whitten, P. (2000) Systematic review of studies of patient satisfaction with telemedicine. *British Medical Journal* 32: 1517–20.
Mair, F., Whitten, P., May, C. and Doolittle, G. C. (2000). Patients' perceptions of a telemedicine specialty clinic. *Journal of Telemedicine and Telecare* 6: 36–40.

Malson, H. (1998) *The Thin Woman: Feminism, Post-structuralism and the Social Psychology of Anorexia Nervosa*. London, Routledge.
Marcuse, H. (1966) *Eros and Civilisation: a Philosophical Inquiry into Freud*. Boston, Beacon Press.
Martin, G. (2000) Stasis in complex artefacts, *in* J. Ziman, (ed.) *Technological Innovation as an Evolutionary Process*. Cambridge, Cambridge University Press.
Martin, P. (2001) Genetic governance: the risks, oversight and regulation of genetic databases in the UK. *New Genetics and Society* 20: 157–83.
Mason, D., Button, G., Lankshear, G. and Coates, S. (2002) Getting real about privacy and surveillance at work, *in* Steve Woolgar (ed.) *Virtual Society? Technology, Cyberbole, Reality*. Oxford, Oxford University Press.
May, C., and Ellis, N. T. (2001) When protocols fail: technical evaluation, biomedical knowledge, and the social production of 'facts' about a telemedicine clinic. *Social Science and Medicine* 53: 989–1002.
May, C., Ellis, N. T., Atkinson, T., Gask, L., Mair, F. S., and Smith, C. (1999) Psychiatry by videophone: a trial service in North West England, *in* P. Kokol, B. Zupan, and J. Stare (eds) *Medical Informatics Europe 99: Bridges of Knowledge*. Amsterdam, IOS Press: 207–10.
May, C., Finch, T., Mair, F. S., and Mort, M. (2005) Towards the wireless patient: chronic illness, scarce care, and technological innovation in the NHS. *Social Science and Medicine* (61) 2005: 1485–94.
May, C., Gask, L., Atkinson, T., Ellis, N., Mair, F. and Esmail, A. (2001) Resisting and promoting new technologies in clinical practice: the case of telepsychiatry. *Social Science and Medicine* 52: 1889–901.
May, C. R., Harrison, R., Finch, T., MacFarlane, A., Mair, F. S. and Wallace, P. (2003) Understanding the normalization of telemedicine services through qualitative evaluation. *Journal of the American Medical Informatics Association* 10: 596–604.
May, C., Mort, M., Mair, F., Ellis, N. T. and Gask, L. (2000) Evaluation of new technologies in health care systems: what's the context? *Health Informatics Journal* 6: 64–8.
May, C., Mort, M., Mair, F. S. and Williams, T. (2001) Factors affecting the adoption of telehealthcare technologies in the United Kingdom: the policy context and the problem of evidence. *Health Informatics Journal* 7: 131–4.
McCartney, M. (2004) The giving game. *Guardian*, 25 May.
McCune, C. A., Al Jader, L. N., May, A., Hayes, S. L., Jackson, H. A. and Worwood, M. (2002) Hereditary haemochromatosis: Only 1 per cent of adult HFE C282Y homozygotes in South Wales have a clinical diagnosis of iron overload. *Human Genetics* 111: 538–43.
McDonnell, S. M., Preston, B. L., Jewell, S. A., Barton, J. C., Edwards, C. Q., Adams, P. C. and Yip, R. (1999) A survey of 2851 patients with Hemochromatosis. *American Journal of Medicine* 106: 619–24.
McGregor, K. K, and Peay, E. R. (1996) The choice of alternative therapy for health care: testing some propositions. *Social Science and Medicine* 43: 1317–27.
McKee, G. K. (1970) Development of total prosthetic replacement to the hip. *Clinical Orthopaedics and Related Research* 72: 85–103.
McPherson, K., Steel, C. M. and Dixon, J. M. (1994) Breast cancer – epidemiology risk factors and genetics. *British Medical Journal* 309: 1003–06.
MDA Medical Device Agency (2002) *A Code of Practice for the Production of Human-derived Therapeutic Products*. London, UK Department of Health.
Mead, N. and Bower, P. (2000) Patient-centredness: a conceptual framework and review of the empirical literature. *Social Science and Medicine* 51: 1087–110.

Melkerson, M. N, and Demian, H. W. (1993) Regulatory perspective for orthopaedic devices, *in* B. F Morrey, *Biological, Material and Mechanical Considerations of Joint Replacement*. New York, Raven Press.

Mendelson, C. (2003) Gentle hugs: Internet listservs as sources of support for women with lupus. *Advances in Nursing Science* 25: 299–306.

Metcalfe, J. S. (1998) *Evolutionary Economics and Creative Destruction*. London, Routledge.

Michie, S., Smith, J. A., Senior, V. and Marteau, T. M. (2003) Understanding why negative genetic test results sometimes fail to reassure. *American Journal of Medical Genetics* 119A: 340–7.

Miller, A. (2002) Telemedicine and doctor–patient communication: a theoretical framework for evaluation. *Journal of Telemedicine and Telecare* 8: 311–18.

Milne C. (2000) The FDA Modernization Act and the FDA: metamorphosis or makeover? *Drug Information Journal* 34: 681–92.

Mody, C. (2001) A little dirt never hurt anyone: Knowledge-making and contamination in materials science. *Social Studies of Science* 31: 7–36.

Mol, A. (2002) Cutting surgeons, walking patients: some complexities involved in comparing, *in* J. Law and A. Mol (eds) *Complexities: Social Studies of Knowledge Practices*. Durham, Duke University Press.

Montbriand, M. (1998) Abandoning biomedicine for alternative therapies: oncology patients' stories. *Cancer Nurse* 21: 36–45.

Morgan, D. (1996) *Family Connections: an Introduction to Family Studies*. Cambridge: Polity.

Morris, Z. (2002) Fears over artificial hips and breasts. *Evening Standard*, 11 April.

Mort, M., and Finch, T. (2005) Principles for telemedicine and telecare: the perspective of a citizens' panel. *Journal of Telemedicine and Telecare* 11. Suppl. 1: 66–8.

Mort, M., May, C. R., and Williams, T. (2003) Remote doctors and absent patients: acting at a distance in telemedicine? *Science, Technology and Human Values* 28: 274–95.

Mulkay, M. (1997) *The Embryo Research Debate: Science and the Politics of Reproduction*. Cambridge, Cambridge University Press.

Mulvey, S. and Wallace, E. M. (2000) Women's knowledge of and attitudes to first and second trimester screening for Down's syndrome. *British Journal of Obstetrics and Gynaecology* 107: 1302–05.

Munro, D. (1949) Rehabilitation stifled by PVA. *Paraplegia News* October: 1, 4, 6.

Munsey, R. R. (1995) Trends and events in FDA regulation of medical devices over the last fifty years. *Food and Drug Law Journal* 163–77.

Murray, D. W., Carr, A. and Bulstrode, C. (1995) Which primary hip replacement? *Journal of Bone and Joint Surgery* 77-B: 520–27.

Myerson, G. (2000) *Donna Haraway and GM Foods*. Cambridge, Cambridge Icon Books.

Najman, J. M. and Levine, S. (1981) Evaluating the impact of medical care and technologies on the quality of life: a review and critique. *Social Science and Medicine* 15: 107–15.

National Audit Office (2002) *NHS Direct in England*. London, The Stationery Office.

National Health Service Executive (1998) *Information for Health: an Information Strategy for the Modern NHS 1998–2001*. London, National Health Service Executive.

Nazroo, J. Y. (1997) *The Health of Britain's Ethnic Minorities*. London, Policy Studies Institute.

Neary, F. (forthcoming) Hip prostheses in local contexts: the comparative roles of surgeons, engineers and manufacturers.

Neary, F. and Pickstone, J. V. (forthcoming) Hip prostheses as technological artefacts: convergence and variety.
Neilson, E. (2003) Securing Good Health for the Whole Population: The Walness 2 Review. Submission by the Royal Pharmaceutical Society of Great Britain, accessed via: http://www.rpsgb.org.uk/pdfs/wanless2sub03.pdf
Nelkin, D. and Tancredi, L. (1994) *Dangerous Diagnostics: the Social Power of Biological Information*. Chicago, University of Chicago Press.
Nettleton, S. (1995) *The Sociology of Health and Illness*. Cambridge, Polity Press.
Nettleton, S., Burrows, R., O'Malley, L. and Watt, I. (2004) Health e-types? An analysis of the everyday use of the Internet for health. *Information, Communication and Society* 7: 4.
Nettleton, S., Burrows, R. and O'Malley, L. (2005) The mundane realities of the everyday lay use of the Internet for health and their consequences for media convergence. *Sociology of Health and Illness* 27 (7): 972–92.
NICE [National Institute for Clinical Excellence] (2000) *Guidance on the Selection of Prostheses for Primary Total Hip Replacement*. Technology Appraisals Guidance No. 2.
NICE (2003) Antenatal care: routine care for the healthy pregnant woman. National Collaborating Centre for Women's and Children's Health Commissioned by the National Institute for Clinical Excellence, London.
Niederau, C., Fischer, R., Purschel, A., Stremmel, W., Haussinger, D. and Strohmeyer, G. (1996) Long-term survival in patients with hereditary haemochromatosis. *Gastroenterology* 110: 1107–19.
Nightingale, P. and Martin, P. A. (2004) The myth of the biotech revolution. *Trends in Biotechnology* 22: 564–9.
Nowlan, A. (2003) Designed around the patient: aspiration or organising principle? Innovative Health Technologies Annual Conference, London, 29–30 October.
Novas, C. and Rose, N. (2000) Genetic risk and the birth of the somatic individual. *Economy and Society* 29: 485–513.
Nowotny H., Scott, P. and Gibbons, M. (2001) *Re-Thinking Science: Knowledge and the Public in an Age of Uncertainty*. London, Routledge.
Nuffield Council on Bioethics (1993) *Genetic Screening: Ethical Issues*. London, Nuffield Council on Bioethics.
O'Cathain, A., Goode, J., Luff, D. Strangleman, T., Hanlon, G. and Greatbatch, D. (2005) Does NHS Direct empower patients? *Social Science and Medicine* 61: 1261–771.
O'Hara, S. (2002) Interview with Nick Watson, Berkeley, 3 October 2002.
Ohras, A., Yelding, D. and Mitchell, J. (1997) *Consumers and their Wheelchairs*. Sheffield, London, The Health Research Institute and the Research Institute for Consumer Affairs.
Oliver, M. (1990) *The Politics of Disablement*. Basingstoke, Macmillan – now Palgrave Macmillan.
Orbach, S. (1993) *Hunger Strike*. London, Penguin.
Oudshoorn, N. (1994) *Beyond the Natural Body: an Archaeology of Sex Hormones*. London, Routledge.
Parr, S. (2004) *Living with Severe Aphasia: the Experience of Communication Impairment after Stroke*. Brighton, Pavilion Press.
Parr, S. (2005) Improving services for stroke and aphasia. *Therapy Weekly* 3 February 2005.
Parr, S., Byng, S. and Gilpin, S. (1997) *Talking about Aphasia*. Buckingham, Open University Press.
Parr, S., Duchan, J. and Pound, C. (eds) (2003) *Aphasia Inside Out*. Maidenhead, Open University Press.

Parsons, T. (1951) *The Social System.* London, Routledge & Kegan Paul.
Patel, A., Carson, D. R. and Patel, P. (1999) Evaluation of an unused 1952 Ridley intraocular lens. *Cataract and Refractive Surgery* 25: 1535–9.
Peltzman, R. L. (1973) *Regulating New Drugs.* Chicago, University of Chicago Press.
Phillips, L. and Nicosoa, A. (1992) Clinical perspectives on wheelchair selection: an overview ... with reflections on past and present of a consumer. *Journal of Rehabilitation Research and Development* Clinical Supplement, 2: 1–7.
Pickering, A. (1995) *The Mangle of Practice: Time, Agency and Science.* Chicago, Chicago University Press.
Pickering, A. (2001) In the Thick of Things. Keynote address at the Conference Taking Nature Seriously, Univ. of Oregon, Eugene, OR, 25–7 Feb. 2001.
PHLS (2000) AIDS/HIV quarterly surveillance tables: UK data to end March 2000, London, Public Health Laboratory Service AIDS Centre (HIV, STD Division, Communicable Disease Surveillance Division).
Pickstone, J. V. (2000a) *Ways of Knowing. A New History of Science, Technology and Medicine.* Manchester, Manchester University Press.
Pickstone, J. V. (2000b) Production, community and consumption: the political economy of twentieth century medicine, *in* R. Cooter and J. Pickstone (eds) *Medicine in the Twentieth Century.* Amsterdam, Harwood Academic Publishers.
Pickstone, J. V. (ed.) (1992) *Medical Innovations in Historical Perspective.* Basingstoke, Macmillan – now Palgrave Macmillan.
Pieters, T. (2003) New molecules, markets and changing regulatory practices, *in* J. Abraham and H. Lawton Smith (eds) *Regulation of the Pharmaceutical Industry.* Basingstoke, Palgrave Macmillan: 146–59.
Pinch, T. and Bijker, W. (1984) The social construction of facts and artefacts: or how the sociology of science and the sociology of technology might benefit each other. *Social Studies of Science* 14: 399–441.
Pollock, A. (2004) *NHS Plc.* London, Verso.
Porter, M. and MacIntyre, S. (1984) What is, must be best: a research note on conservative and deferential responses to antenatal care provision. *Social Science and Medicine* 19: 1197–200.
Prentice, D. A. (2003) *Scientific Basis for Cell and Tissue Therapy.* Presentation to Public Hearing on the proposed directive on Quality and Safety of Human Tissues and Cells, Brussels, 29 January 2003. http://www.eutop.de/ct/
Press, N. A. and Browner, C. H. (1993) 'Collective fictions': similarities in reasons for accepting maternal serum screening among women of diverse ethnic and social class backgrounds. *Fetal Diagnosis and Therapy* 8: 97–106.
Press, N., Browner, C. H, Tran, D., Morton, G. and Le Master, B. (1997) Provisional normalcy and 'perfect babies': pregnant women's attitudes toward disability in the context of prenatal testing, *in* S. Franklin and H. Ragone (eds) *Reproducing Reproduction, Kinship, Power and Technological Innovation.* Pennsylvania, University of Pennsylvania Press.
Preston, P. (2001) Telemedicine in the NHS (Presentation given to the British Association of Dermatologists, Symposium on Teledermatology) Royal College of Physicians of London, June 2001.
Prins, B. (1995) The ethics of hybrid subjects: feminist constructivism according to Donna Haraway. *Science, Technology and Human Values* 20: 352–67.
Prior, L., Wood, F., Gray, J., Pill, R. and Hughes, D. (2002) Making risk visible: the role of images in the assessment of genetic risk. *Health, Risk and Society* 4: 242–58.

Rabinow, P. (1992) Artificiality and enlightenment: from socio-biology to biosociality, *in* J. Crary and S. Kwinter (eds) *Zone 6: Incorporations.* Cambridge, MIT Press: 91–112.
Rapp, R. (1999) *Testing Women, Testing the Fetus: the Social Impact of Amniocentesis in America.* New York, Routledge.
Rapp, R. (2003) Cell life and death, child life and death, *in* S. Franklin and M. Lock (eds), *Remaking Life and Death: Towards an Anthropology of Biomedicine.* Santa Fe, NM: School of American Research Press (Oxford: James Currey Ltd.): 129–64.
Rapp, R., Heath, D. and Taussig, K. (2001) Genealogical dis-ease: where heredity, abnormality, biomedical explanation, and family responsibility meet, *in* S. Franklin and S. McKinnon (eds) *Relative Values: Reconfiguring Kinship Studies.* Durham, NC, Duke University Press: 384–412.
Rappert, B. and Brown, N. (2000) Putting the future in its place: comparing innovation moments in genetic diagnostics and telemedicine. *New Genetics and Society* 19: 49–75.
Rees, L. R., and Weil, A. (2001) Integrated Medicine. *BMJ* 322: 119–20.
Reichert, J. M. (2003) Trends in development times for new therapeutics in the US. *NRDD* 2: 695–70.
Reilly, D. (2001) CAM in Europe: Reflections and trends. Article prepared for 2nd International Joint Symposium, Seoul, Korea.
Reilly, D. (date unknown) Creating therapeutic consultations: some reflections. Unpublished lecture transcript.
Reimann, A. (2003) Statement provided at the hearing on the Patient view to the proposed directive on Quality and Safety of Human Tissues and Cells, Brussels, 29 January, 2003. http://www.eutop.de/ct/
Renner, B. (2004) Biased reasoning: adaptive responses to health risk feedback. *Personality and Social Psychology Bulletin* 30: 384–96.
Rheinberger, H. J. (2000) Beyond nature and culture: modes of reasoning in the age of molecular biology and medicine, *in* M. Lock, A. Young and A. Cambrosio (eds) *Living and Working with New Medical Technologies.* Cambridge, Cambridge University Press.
Ridley, H. (1946) Proceedings of the Royal Society of Medicine, Vol. 39.
Ridley, H. (1951) Intra-ocular acrylic lenses. *Transactions of the Ophthalmological Society of the UK* 71: 617–21.
Ridley, H. (1952) Further observations on intra-ocular acrylic lenses. *Transactions of the Ophthalmological Society of the UK* 72: 511–14.
Ridley, H. (1958) Cataract surgery with particular reference to intra-ocular implants of various types. *Transactions of the Ophthalmological Society of the UK* 78: 585–92.
Ring, P. A. (1968) Complete replacement arthroplasty of the hip by the Ring prosthesis. *British Journal of Bone and Joint Surgery* 50: 720–31.
Roberts, E. V. (1994) The UC Berkeley Years: First Student Resident at Cowell Hospital, 1962, an oral history conducted in 1994 by Susan O'Hara in University of California's Cowell Hospital Residence Program for Physically Disabled Students, 1962–1975: Catalyst for Berkeley's Independent Living Movement, Regional Oral History Office, The Bancroft Library, Berkeley, University of California.
Roberts, R. and Rigby, M. (1998) The need for a holistic view of telemedicine, focused on patients and society as prime stakeholders, *in* B. Cesnik, A. McCray and J. R. Scherrer (eds) *Medinfo.* Amsterdam, IOS Press, pp. 1204–08.
Rose, D. and Blume, S. (2003) Citizens as users of technology: an exploratory study of vaccines and vaccination, *in* N. Oudshoon and T. Pinch (eds) *How Users Matter: the Co-construction of Users and Technology.* Cambridge, MIT Press.

Rose, N. (2001) The politics of life itself theory. *Culture and Society* 18: 1–30.
Rose, N. (2004) Becoming neurochemical selves, *in* N. Stehr (ed.) *Between Commerce and Civil Society: Biotechnology*. New Brunswick, NJ, Transaction Publishers: 89–126.
Rosenbrock, R., Dubois-Arber, F., Moers, M., Pinell, P., Schaeffer, D. and Setbon, M. (2000) The normalisation of AIDS in Western European countries. *Social Science and Medicine* 50: 1607–29.
Rosengarten, M., Imrie, J., Flowers, P., Davis, M. and Hart, G. (2004) After the euphoria: HIV medical technologies from the perspective of clinicians. *Sociology of Health and Illness* 26: 575–96.
Royal College of Surgeons (2001) *3M Capital Hip Ststem: the Lesson learned from the Investigation*. London.
Royal Pharmaceutical Society of Great Britain (2004) 'Perspectives on the Expert Patient,' Presentations from a seminar held at RPSGB 19 May 2003, accessed via: http://www.rpsgb. org.uk/pdfs/exptpatsemrept.pdf.
Royal Society (1992) *Risk: Analysis, Perception and Management*. London, Royal Society.
Royal Society (1997) *Science, Policy and Risk*. London, Royal Society.
Rowe, R. E., Garcia, J. and Davidson, L. L., (2004) Social and ethnic inequalities in the offer and uptake of prenatal screening and diagnosis in the UK: a systematic review. *Public Health* 118: 177–89.
Sabin, C., Hill, T., Lampe, F., Mathias, R., Bhagani, S., Gilson, R., Youle, M., Johnson, M., Fisher, M., Scullard, G., Easterbrook, P., Gazzard, B. and Phillips, A. (2005) Treatment exhaustion of highly active antiretroviral therapy (HAART) among individuals infected with HIV in the Unitied Kingdom: multicentre cohort study. *BMJ* doi:10.1136/bmj.38369.669850.8F.
Sackett, D. L., and Rosenberg, W. M. (1995) The need for evidence-based medicine. *Journal of the Royal Society of Medicine* 88: 620–4.
Sackett, D. L., Rosenberg, W. M., and Gray, J. A., et al. (1996) Evidence based medicine: what it is and what it isn't. *British Medical Journal* 312: 71–2.
Salter, B. and Jones, M. (2002) Human genetic technologies: European governance and the politics of bioethics. *Nature Reviews Genetics* 3: 808–14.
Sandall, J., Grellier, R. and Ahmed, S. (2003) Prenatal screening and diagnosis in a multi-cultural, multi-ethnic society, *in* L. Abramsky and J. Chapple (eds) *Prenatal Diagnosis*. London, Chapman and Hall: 83–97.
Sandall, J., Pitson, L., Lewando Hundt, G., Williams, C., Spencer, K. and Heyman, B. 'Going with the flow' routinisation and constraints on informed decision-making and non directiveness in a one-stop clinic offering first trimester prenatal screening for Down's Syndrome: a cross sectional survey of women's experiences and views (under review).
Sandman, L. (2004) *A Good Death: On the Value of Death and Dying*. Buckingham, Open University Press.
Sarmiento, A. (2003) *Bare Bones: a Surgeon's Tale*. New York, Prometheus Books.
Savolain, R. and Kari, J. (2004) Conceptions of the Internet in everyday life information seeking. *Journal of Information Science* 30: 219–26.
Scales, J. T. and Wilson J. N. (1969) Some aspects of the development of the Stanmore total hip joint prosthesis. *Reconstruction Surgery and Traumatology*, 11: 20–39.
Schlich, T. (2002) *Surgery, Science and Industry. a Revolution in Fracture Care, 1950s–1990s*. Basingstoke, Palgrave Macmillan.
Schweitzer, S. O., Schweitzer, M. E. and Sourty-Le Guellec, M-J. (1996) Is there a US drug lag? The timing of new pharmaceutical approvals in the G-7 countries and Switzerland. *Medical Care Research and Review* 53: 162–78.

SCMPMD (2001a) *Scientific Committee on Medicinal Products and Medical Devices. Opinion on the State of the Art Concerning Tissue Engineering.* http://europa.eu.int/comm/food/fs/sc/scmp/out37_en.pdf. Accessed 18/03/03

SCMPMD (2001b) *Opinion on the State of the Art Concerning Xenotransplantation.* SANCO/SCMPMD/2001/0002. http://europa.eu.int/comm/health/ph_risk/committees/scmp/documents/out38_en.pdf (accessed 12/04).

Seale, C. (2006) New directions for critical Internet health studies: representing cancer experiences on the Web. *Sociology of Health And Illness* 27(4): 515–40.

Seymour, J. E. (2002) Artificial feeding at the end of life: older people's understandings, in C. Gastmans (ed.) *Between Technology and Humanity: the Impact of New Technologies on Health Care Ethics*. Brussels, Leuven University Press.

Seymour, J. E., Clark, D., Gott, M., Bellamy, G. and Ahmedzai, S. (2002) Good deaths, bad deaths: older people's assessments of risks and benefits in the use of morphine and terminal sedation in end of life care. *Health, Risk and Society* 4: 287–304.

Seymour, J. E., Gott, M., Bellamy, G., Clark, D. and Ahmedzai, S. (2004) Planning for the end of life: the views of older people about advance statements. *Social Science and Medicine* 59: 57–68.

Sharma, U. (1992) *Complementary Medicine Today: Practitioners and Patients*. London, Routledge.

Sharp, L. A. (2002) Bodies, boundaries, and territorial disputes: investigating the murky realm of scientific authority. *Medical Anthropology* 21: 369–79.

Shaw, J. and Baker, M. (2004) Expert patient: dream or nightmare? *British Medical Journal* 328: 723–4.

Shelley, R (1997) *Anorectics on Anorexia*. London, Jessica Kingsley Publishers.

Shilling, C. (2002) Culture, the 'sick role' and the consumption of health. *British Journal of Sociology* 53(4): 621–38.

Sicotte, C. and Lehoux, P. (2003) Teleconsultation: rejected and emerging uses. *Methods of Information in Medicine* 42: 451–7.

Siegrist, J. and Junge, A. (1989) Conceptual and methodological problems in research on the quality of life in clinical medicine. *Social Science and Medicine* 29: 463–8.

Simmons Mackie, N. (2000) Social approaches to the management of aphasia, in L. Worrall and C. Frattali (eds) *Neurogenic Communication Disorders: a Functional Approach*. New York, Thieme.

Skolbekken, J. A. (1995) The risk epidemic in medical journals. *Social Science and Medicine* 40: 291–305.

Slevin, J. (2000) *The Internet and Society*. Oxford, Polity Press.

Smith, R. (2002) The future of medical education: speculation and possible implications 5.7.2002, www.bmj.com/talks

Sparber, A., Bauer, L. and Curt, G. (2000) Use of complementary medicine by adult patients participating in cancer clinical trials. *Oncology Nursing Forum* 27: 623–30.

Spencer, K. and Power, M. et al. (2003) Screening for chromosomal abnormalities in the first trimester using ultrasound and maternal serum biochemistry in a one-stop clinic: a review of three years prospective experience. *British Journal of Obstetrics and Gynaecology* 110: 281–6.

Stacey, M. (1988) *The Sociology of Health and Healing*. London, Unwin Hyman.

Stanberry, B. (2001) Legal, ethical and risk issues in telemedicine. Computer methods and programs. *Biomedicine* 64: 225–33.

Stanton, J. (2002) *Innovation in Health and Medicine: Diffusion and Resistance in the Twentieth Century*. London, Routledge.

Stapleton, H., Kirkham, M. and Thomas, G. (2002) Qualitative study of evidence based leaflets in maternity care. *British Medical Journal* 324: 639–43.

Stiker, H. J. (1999) *The History of Disability*. Michigan, Michigan University Press.
Stimpson, G. and Webb, B. (1975) *Going to See the Doctor: the Consultation Process in General Practice*. London, Routledge and Kegan Paul.
Stone, D. (1984) *The Disabled State*. Philadelphia, Temple University Press.
Subcommittee on Oversight and Investigations (1983) Medical device regulation: the FDA's neglected child. Washington.
Suchman, L. A. (2002) Practice-based design of information systems: notes from the hyperdeveloped world. *The Information Society* 18: 139–44.
Swain, J. and French, S. (2000) Towards an affirmation model of disability. *Disability & Society* 15: 569–82.
Swinburn, W. R. (ed.) (1983) *Wrightington Hospital 1933–1983*. Wrightington Hospital.
Sygall, S. (1998) 'Cofounder and Director of Berkeley Outreach Recreation Program and Mobility International USA, Advocate for Women's Issues', an oral history conducted in 1998 by Kathryn Cowan, Regional Oral History Office, The Bancroft Library, University of California, Berkeley, 2000.
Tang, S. Y. S. and Anderson, J. M. (1999) Human agency and the process of healing: lessons learned from women living with chronic illness – 're-writing the expert'. *Nursing Inquiry* 6, 83–93.
Thompson, F. R. (1954) Two and a half years' experience with a Vitallium intramedullary hip prosthesis. *Journal of Bone and Joint Surgery* 36-A: 489–500.
Thorne, S. E., Ternulf Nyhlin, K. and Paterson, B. L. (2000) Attitudes toward patient expertise in chronic illness. *International Journal of Nursing Studies* 37: 303–11.
Treichler, P. (1999) *How to have Theory in an Epidemic: Cultural Chronicles of AIDS*. Durham, Duke University Press.
Tuckett, D., Boulton, M., Olson, C. and Williams, A. (1985) *Meetings between Experts*. London, Tavistock.
Tufts Center for the Study of Drug Development (2004) *Outlook 2004*. Boston, MA, Tufts CSDD.
TUFTS (2006) *Outlook Report*, Tufts Centre for Drug Discovery, Boston, MA.
Turner, B. S. (1996) *The Body and Society*. London, Sage.
Turner, N. (2004) Pricing climate heats up in US and Europe. *Pharmaceutical Executive* (July).
Turner, V. (1967) Betwixt and between: the liminal period in rites de passage, *in The Forest of Symbols: Aspects of Ndembu Ritual*. Ithaca: Cornell University Press: 93–111.
Tutton, R. (2002) Gift relationships in genetics research. *Science as Culture* 11: 523–42.
Tutton, R., Kerr, A. and Cunningham-Burley, S. (2005) Myriad Stories: Constructing expertise and citizenship in discussions of the new genetics. In M. Leach, J. Scoones and B. Wynne (eds) *Science and Citizens: Globalisation and the Challenge of Engagement*. London, Zed Press, pp. 101–12.
Umberson, D. (1992) Gender, marital status and the social control of health behaviour. *Social Science & Medicine* 34: 907–17.
United Kingdom Xenotransplantation Interim Regulatory Authority (2003) Fifth Annual Report. London, UK Department of Health.
United States General Accounting Office Health Education and Human Services Division Report to Subcommittee on Oversight and Investigations (1998) Medical Devices: FDA can improve Oversight of Tracking and Recall Systems, Washington.
UPIAS (1979) *Fundamental Principles of Disability*. London.
Urquhart, C., Currell, R., Lewis, R. and Wainwright, P. (1999) Evaluating telemedicine: indications from a systematic review: 186–191. HC 99, Current perspectives in healthcare computing. Harrogate, 22–4 March.

Urwin, R. E., Bennetts, B. and Wilcken, B. et al. (2002) Anorexia nervosa (restrictive subtype) is associated with a polymorphism in the novel norepinephrine transporter gene promoter polymorphic region. *Molecular Psychiatry* 7: 652–7.

US Congress Office of Technology Assessment (1987) *Technology-Dependent Children: Hospital v. Home Care – A Technical Memorandum, OTA-TM-H-38*. US Government Printing Office, Washington DC.

US Senate (1997) *Food and Drug Administration Modernization Act: Report together with additional and minority reviews to accompany S.830*.

Vasen, H. F. A., Mecklin, J. P., Merakhan, P. and Lynch, H. T. (1991) The international collaborative group on hereditary non-polyposis colorectal cancer. *Diseases of the Colon and Rectum* 34: 424–5.

Vos, R. (1991) *Drugs looking for diseases*. Lancaster, Kluwer.

Wadsworth, M. E. J., Butterfield, W. and Blaney, R. (1971) Perception of illness and use of services in an urban community. *Health and Sickness: the Choice of Treatment*. London: Tavistock.

Wajcman, J. (2002) Addressing technological change: the challenge to social theory. *Current Sociology* 50: 347–63.

Waldby, C. (1997) The body and the digital archive: the visible human project and the computerisation of medicine. *Health* 1: 77–90.

Waldby, C. (2002) Stem cells, tissue cultures and the production of biovalue. *Health* 6: 305–23.

Wardell, W. (1973) Introduction of new therapeutic drugs in the US and Great Britain: an international comparison. *Clinical Pharmacology and Therapeutics* 14: 773–90.

Watson, N. and Woods, B. (2005) No wheelchairs beyond this point: a historical examination of wheelchair access in the twentieth century in Britain and America. *Social Policy & Society* 4: 97–105.

Watts, M. S. (1970) Ecological health and quality of life now and forevermore. *California Medicine* 113: 55–7.

Waugh, W. (1990) *John Charnley: the Man and the Hip*. London, Springer-Verlag.

Webster, A. (2000) The innovative health system: implications for social science research. IHT Programme Launch, York.

Webster, A. (2002). Innovative health technologies and the social: redefining health, medicine and the Body. *Current Sociology* 50: 443–57.

Weiss, G. (2000) *Body Images: Embodiment as Intercorporeality*. London, Routledge.

Wheale, A. (2001) Science advice, democratic responsiveness and public policy. *Science and Public Policy* 26: 413–21.

Whitten, P. and Collins, B. (1997) The diffusion of telemedicine: communicating an innovation. *Science Communication* 19: 21–40.

Whitten, P., Mair, F., Haycox, A., May, C., Williams, T., and Helmich, S. (2002) Systematic review of cost effectiveness studies of telemedicine interventions. *British Medical Journal* 324: 1434–7.

Wiles, P. W. (1958) The surgery of the osteoarthritic hip. *British Journal of Surgery* 45: 488–97.

Wilkin, T. and Devendra, D. et al. (2001) Education and debate: bone densitometry is not a good predictor of hip fracture. *BMJ* 323: 795–9.

Williams, C., Alderson, P. and Farsides, B. (2002a) 'Drawing the line' in prenatal screening and testing: health practitioners discussions. *Health Risk and Society* 4: 61–75.

Williams, C., Alderson, P. and Farsides, B. (2002b) Too many choices? Hospital and community staff reflect on the future of prenatal screening. *Social Science and Medicine* 55: 743–53.

Williams, C., Sandall, J., Lewando Hundt, G., Spencer K., Heyman B., Grellier, R., Heyman, B. and Spencer, K. (2005) Women as 'moral pioneers'? Experiences of first trimester nuchal translucency screening. *Social Science and Medicine* 61: 1983–92.

Williams, P., Huntington, P. and Nicholas, D. (2003) Health information on the Internet: a qualitative study of NHS Direct online users. *Aslib Proceedings* 55: 304–12.

Wilson, P. M. (2001) A policy analysis of the Expert Patient in the United Kingdom: self-care as an expression of pastoral power? *Health and Social Care in the Community* 9: 134–42.

Winner, L. (1980) Do artifacts have politics? *Daedalus* 109: 121–36.

Wolpe, P. R. and McGee, G. (2001) Expert bioethics as professional discourse: the case of stem cells, *in* S. Holland, K. Lebacqz and L. Zoloth (eds) *The human Embryonic Stem Cell Debate: Science, Ethics and Public Policy.* Cambridge, MIT Press: 185–96.

Wood, F., Prior, L. and Gray, J. (2002) Translations of risk: decision making in a cancer genetics clinic. *Health, Risk and Society* 5: 183–98.

Woods, B., and Watson, N. (2003) A short history of powered wheelchairs. *Assistive Technology* 15: 164–80.

Woods, B. and Watson, N. (2004a) In pursuit of standardization: the British Ministry of Health's model 8F wheelchair 1948–1962. *Technology and Culture* 45: 540–68.

Woods, B. and Watson, N. (2004b) A glimpse at the social and technological history of wheelchairs. *International Journal of Therapy and Rehabilitation* 11: 407–10.

Woolgar, S. (1991) Configuring the user: the case of usability trials, *in* J. Law (ed.) *A Sociology of Monsters.* London, Routledge.

Wootton, R. (1997) *Medicine on the Superhighway.* Telemed 97. Conference proceedings 26–7 November.

World Wide Web Consortium (1999) Web accessibility guidelines 1.0. Retrieved 3 December 2002, from http://www.w3org/TR/WCAG10/

Worwood, M. (1999) Inborn errors of metabolism: iron. *British Medical Bulletin* 55: 556–67.

Writing Group for the Women's Health Initiative Investigators (2002) Risks and benefits of estrogen plus progestin in healthy postmenopausal women. *JAMA* 288: 321–33.

Wyatt, J. C. (1996) Commentary: telemedicine trials – clinical pull or technology push? *British Medical Journal* 313: 1380–1.

Young, M. (1988) *The Metronomic Society: Natural Rhythms and Human Timetables.* London, Thames and Hudson.

Young, M. and Schuller, T. (eds) (1988) *The Rhythms of Society.* London, Routledge.

Ziebland, S. (2004) The importance of being an expert: the quest for cancer information on the Internet. *Social Science & Medicine* 59: 1783–93.

Zola, I. K. (1973) Pathways to the doctor – from person to patient. *Social Science and Medicine* 7: 677–89.

Author Index

Aanestad, M. 86
Abraham, J. 177, 181, 182, 183, 184, 186, 188
Abraham, S. 102
Abramsky, L. 38
Abramson, S. 162
Adam, B. 137
Adamson, A. 141
Adkins, L. 106
Akrich, M. 164, 239
Allen, K. 13
Ammenwerth, E. 84
Anderson, J. M. 99, 155
Anderson, T. 163
Andersson, F. 184
Angell, M. 224
Annandale, E. 133
Anon 183, 184
Ansell Pearson, K. 196
Apple 149, 151
Armstrong, D. 2, 226, 234
Ashworth, M. 98
Astin, J. A. 227
Atkinson 86
Audit Commission 162

Baker, M. 99
Ball, M. J. 84
Barad, K. 209
Barlow, J. 86
Barnartt, S. 167
Barnes, C. 163
Barry, A. 197
Bayer, S. 86
Beck, U. 22, 104, 106, 131, 132
Beck-Gernsheim, E. 61, 104
Berg, M. 2, 86
Bessell, T. L. 98
Beutler, E. 12, 13
Bharadwaj, A. 14
Bijker, W. 113, 161, 209
Billings, C. F. 167
Binkhorst, C. D. 149
Block, M. 226

Blume, S. 155, 165
Blumenthal, D. 60, 66
BMA 62, 141
Bognor, W. C. 188
Booth, B. 189
Bordo, S. 102
Bothwell, T. H. 12
Bourke, J. 162
Bower, P. 226
Boyd, E. A. 226
Bracken, P. 113
Brett, A. S. 17
Briggs, E. 166
Broom, A. F. 61
Brown, N. 26, 82, 85, 194, 195, 232
Browner, 25, 27
Bruch, H. 102
Brumberg, J. J. 101
Bryan, L. 226
Buanes, T. 86
Bud, R. 196
Burke, W. 13
Burrows, R. 72
Bury, M. 132
Byng, S. 173

Cahill, S. E. 164
Caldwell, D. 234
Callon, M. 209
Cane, J. 165
Carter, S. 196
Charatan, F. 61
Charles River Associates 178, 179, 185
Chatwin, J. 227
Chilaka, V. N. 25
Chisholm, K. 102
Clark, E. A. 13, 133
Clinical Standards Advisory Group 225
Clough, K. 85
CMR International 179
Collins, H. M. 8, 85, 227

Author Index

Conrad, P. 62, 68
Corbin, J. 63
Corby, P. 163
Cornford, T. 85
Coulehan, J. L. 226
Cranney, A. 72
Crawford, R. 17
Cullen, J. 115
Cunningham-Burley, S. 42, 53
Currell, R. 86
Curry, R. 86

Daly, K. J. 137
Davis, M. 110, 184
Davis-Floyd, R. W. 27, 133
Davison, C. 17
Deleuze, G. 196
Denis, J. L. 86
Department of Health 13, 61, 85, 97, 168, 199
Devendra, D. 72
Di Masi, J. 185
Dias, K. 102
Dickinson, D. 226
Doolittle, G. C. 86
Douglas, M. 195, 196
Doward, J. 102
Drews, J. 187

Eastwood, H. 224
Eder, K. 131
Edwards, A. 14
Edwin, B. 86
Eedy, D. J. 91
Eggleston, R. 164
Eisenberg, D. M. 224
Ellershaw, J. 140
Ellis, N. T. 86
Elman, R. 173
Elwyn, G. 226
EMEA 185
Epstein, S. 105
Erni, J. 105
Ernst, E. 228
Esmail, A. 86
EU Commission 205, 213, 218
EuropaBio 205
European Council 215
European Parliament 219, 221
Evans, J. 8, 214

Farquhar, J. 40
Faulkner, A. 194, 196
FDA 179, 185, 192, 198
Featherstone, M. 143
Feenberg, A. 113, 114
Finch, T. 85, 88, 91, 95
Finkler, K. 17, 23, 48
Fitzgerald, F. T. 224
Fitzpatrick, R. 159, 235
Flower, R. 98
Flowers, P. 105, 110
Forde, O. H. 14
Foucault, M. 2, 112
Fox, N. 98, 194
Frank, A. 170, 171
Franklin, S. 42, 48, 49, 50, 53, 194, 195
Freeman, C. 181
Freidson, E. 58
French, S. 163
Furnham, V. C. 227
Fuss, M. 167

Gabe, J. 14, 196
Galambos, L. 190
Gann, B. 85
Gannon, L. 71
Gask, L. 86, 88
Gastaldo, D. 100
Gelijns, A. C. 155
Gibbons, M. 196
Giddens, A. 3, 22, 126, 137
Gifford, S. M. 14, 15, 17
Gilbert, D. 98
Gillett, J. 61
Ginsburg, F. 40
Glendinning, C. 136
Goldsmith, S. 167
Goode, J. 63, 65
Goodman, J. 187
Gordon, R. A. 102
Gott, M. 144
Gottweis, H. 213
Graber, M. A. 61
Grabowski, H. 177
Gramsci, A. 114
Gray, J. 16, 233
Green, E. 74, 75, 80, 212
Green, J. 25
Green, R. M. 212

Greene, K. R. 133
Greenhalgh, T. 226
Gremillion, H. 102
Griffiths, F. E. 71, 75
Griffiths, J. 134
Grimes, C. A. 167
Grogan, S. 102
Guattari, F. 196
Gutman, E. 167
Gwyn, R. 226

Habermas, H. 112
Hacking, I. 14
Hailey, D. 86
Hallam, E. 143
Hamilton, R. C. 148
Hancher, L. 180
Hanlon, G. 62, 63, 65, 69
Hannay, D. R. 11
Haraway, D. 196
Hardey, M. 61, 72
Harrison, R. 85
Hart, A. 72
Haux, R. 84
Haycox, A. 86
Heath, C. 40
Heaton, J. 137
Hedgecoe, A. 196
Heidegger, M. 112
Helmich, S. 86
Henderson, R. 186
Henwood, F. 61, 72, 98, 99
Hepworth, M. 143
Heritage, J. 226
Herzog, W. 84
Hewitt, A. 173
Hickey, A. M. 235
Hobson, D. 166
Hoberman, M. 162
Hochschild, A. 66
Hockenberry, J. 164
Hoffman, B. 5
Hogle, L. 197, 209
Holohan, A. 63
Horton, R. 60, 66
House of Lords 225
Hudson-Jones, A. 226
Hughes, K. 168
Hughes, T. P. 210

Hunt, D. L. 134
Hurwitz, B. 226

Illich, I. 140
Illman, J. 97

Jacob, F. 23
Jacobson, J. 165
Jaffe, N. S. 158
Jardine, I. 85
Jasanoff, S. 47, 214
Jaspers, K. 113
Jevons, F. 181
Jones, K. 71
Jones, M. 214

Kaitin, K. I. 185
Kamenetz, H. 161
Kaplan, B. 134, 136
Kari, J. 61, 63
Kaufman, H. E. 147
Keith, R. D. F. 133
Kelman, C. D. 150
Kendall, L. 85
Kennedy, J. 27
Kerr, A. 26, 40, 53
Kessler, D. A. 183
Khosa, J. 169
Klecun-Dabrowska, E. 85
Klenerman, L. 151
Klinge, I. 76
Knaup, P. 84
Kneller, R. 191
Knorr-Cetina, K. 209
Kong, Y. 204
Konrad, M. 48, 49

Lacroix, A. 86
Lash, S. 63, 69
Lask, B. 226
Last, J. M. 11, 58, 59
Latour, B. 195, 196, 239
Launer, J. 226
Lauritzen, S. O. 74
Lawrence, G. 155
Lefebvre, R. C. 17
Lehoux, P. 85, 86
Lenaghan, J. 85
Lerner, I. J. 224

Levine, S. 235
Lewando Hundt, G. 27
Lewis, G. 177, 183
Lewis, R. 86
Lillis, M. J. 84
Lindsay, C. 102
Linebarger, E. L. 147, 151
Ling, R. 153
Lippman, A. 27
Llewellyn-Jones, D. 102
Loader, B. 61
Lock, M. 40, 194
Luck, J. 103
Lupton, D. 2, 98, 144
Lynn, J. 141

MacFarlane, A. 85
MacKenzie, D. 86
Mackie, L. 169, 173
MacPhail, A. P. 12
MacSween, M. 102
Mair, L. 85, 86, 88
Majone, G. 180, 181
Malson, H. 102
Marcuse, H. 114
Martin, G. 157, 196
Mason, D. 134
May, C. 85, 86, 88, 89
McCartney, M. 159
McCune, C. A. 15
McDonnell, S. M. 16
McGee, G. 213
McGregor, K. K. 224
McKee, G. K. 152
McPherson, K. 13
Mead, N. 226
Mendelson, C. 109
Metcalfe, S. 146
Michael, M. 194, 195
Michie, S. 20
Mody, C. 196
Mol, A. 2, 5
Montbriand, M. 227
Moore, D. 200
Moran, M. 180
Morgan, D. 66
Morris, Z. 159
Mort, M. 85, 86, 88, 91, 95
Mulkay, M. 212

Mulvey, S. 25
Munro, D. 162
Murray, D. W. 159
Myerson, G. 196

Najman, J. M. 235
National Audit Office 62
Neary, F. 154
Neilson, E. 97
Nelkin, D. 23
Nettleton, S. 61, 62, 63, 68, 69
NICE 26, 159
Nicosoa 162
Niederau, C. 12, 16
Novas, C. 23, 50
Nowlan, A. 58
Nowotny, H. 196

O'Cathain, A. 62, 69
O'Hara, S. 166, 167
Ohinmaa, A. 86
Ohras, A. 162
Oliver, M. 163
Orbach, S. 102
Orsenigo, L. 186

Parr, S. 169, 174
Parsons, T. 17, 58
Peay, E. R. 224
Peltzman, R. L. 187
Phillips, L. 162
PHLS 104
Pickering, A. 3
Pickstone, J. 2, 58, 146, 154, 155, 156
Pinch, T. 113, 161
Pisano, G. 186
Pollock, A. 160
Press, N. A. 25, 27
Preston, P. 95
Prins, B. 196
Prior, L. 16

Rabinow, P. 53, 195
Rapp, R. 27, 40, 48, 50
Rappert, B. 85
Reed, T. 186
Rees, L. R. 226
Reichart, J. M. 190
Reilly, D. 102, 224, 227

Renner, B.   20
Rheinberger, H. J.   204
Ridley, H.   147
Rigby, M.   86
Roberts, C.   42, 49, 50, 53
Roberts, E.   166
Roine, R.   86
Rose, N.   23, 40, 50, 145, 165
Rosenberg, N.   155, 233
Rosenbrock, R.   107
Rosengarten, M.   105
Rotnes, J. S.   86
Royal College of Surgeons   154
Royal Society   14
RPSGB   98

Sabin, C.   105, 106
Sachs, L.   74
Sackett, D. L.   233
Salter, B.   214
Sandall, J.   27
Sandman, L.   140
Sarmiento, A.   159
Saunders, D.   27
Savolain, R.   61
Scales, J. T.   152
Schlich, T.   153, 157
Schuller, T.   137
Schweitzer, S. O.   184
SCMPMD   202
Scotch, R.   167
Seymour, J.   142, 143, 144
Shakespeare, T.   26
Sharma, U.   224
Sharp, L. A.   197, 209
Shaw, J.   99
Shelley, R.   101, 102
Shilling, C.   62
Sicotte, C.   85, 86
Siegrist, J.   235
Simmons Mackie, N.   169, 173
Skolbekken, J. A.   14
Slevi, J.   61
Smith, R.   59, 86
Spencer, K.   27
Stacey, M.   58, 62, 65, 70
Stanberry, B.   86
Stanton, J.   155
Stapleton, H.   27
Stiker, H. J.   162

Stimpson, G.   58, 62, 70
Stone, D.   162
Strauss, A.   62, 63
Sturchio, J. L.   190
Suchman, L.   96
Swain, J.   183
Swazey, J. P.   194
Sygall, S.   163

Tancredi, L.   23
Tang, S. Y. S.   99
Taussig, K.   40
Then, K. Y.   204
Thomas, P.   113
Thompson, F. R.   74, 80
Thorne, S. E.   99, 100
Timmermann, S.   155
Treichler, P.   105
Tuckett, D.   62
Tufts Center for the Study of Drug Development   185, 191
Turner, B.   2, 17, 140
Turner, N.   179
Tutton, R.   42, 53

Umberson, D.   65
Urquhart, C.   86
Urwin, R. E.   102
US Senate   185

Vasen, H. F. A.   15
Vos, R.   181, 187

Wadsworth, M. E. J.   58
Wainwright, S.   86
Wajcman, J.   84, 86
Waldby, C.   2, 195, 209, 210
Wallace, E. M.   25
Wallace, P.   85
Ward, C.   140
Wardell, W.   177
Watson, N.   164, 166, 167
Watts, M. S.   235
Waugh, W.   152
Webb, B.   58, 62, 70
Webster, A.   2, 26, 77, 81, 84, 194, 232
Weil, A.   226
Weiss, G.   210
Whitten, P.   85, 86

Wilkin, T. 72
Williams, C. 26, 61, 86, 88
Williamson, R. 13
Wilson, P. M. 99, 152
Winner, L. 161
Wolpe, P. R. 213
Wood, F. 16
Woods, B. 164, 166, 167
Woolgar, S. 164

Wootton, R. 85, 91
Worwood, M. 12
Wyatt, S. 72, 85

Young, M. 137

Zemmel, R. 189
Ziebland, S. 61
Zola, I. 57, 58, 69

Subject Index

Anorexia 97 *et passim*
Aphasia 168–72, 173

Biobanks 2, 40, 42, 43
Biotechnology 190–1, 209, 215
Brain imaging 115–18, 122–3, 124

Cancer 11 *et passim*
  and complementary medicine 228–31
Clinical iceberg 58, 59
Clinical trials 181–2, 187, 233
Cochlear implant 155
Collaborative knowledge systems 5, 115–16, 118, 119, 122–4
Complementary medicine 58–9, 224–31
Computerised decision support systems 132–6
CRT scanners 155

Death 131
  natural death 140–4
  and technological mediation 239
Digital interactive TV 120
Disability 161 *et passim*
Down's syndrome 25 *et passim*
Drug approvals 183–6

E-health 61, 65, 85, 88
Electronic patient record 2, 87
Ethnicity 33
Eugenics 27
Evidence-based medicine 233, 239

Fetal well-being 132–6
Food and Drug Administration 158, 177, 178–9, 183–5, 192, 198, 199
Framework Programme Six 215–16

Gendered health care 65
Genetic citizenship 40

Genetics 2, 11 *et passim*
  Genetic iceberg 11
  and information 42, 53
  lay perspectives 44–7
  and testing 33
Genomics 191, 192
Governance 87, 121
  and hybrid tissue 194 *et passim*, 213
  and telemedicine 92–4

Health technology evaluation 7–8, 145, 232 *et passim*,
HFEA 218
Hip replacement 146, 151–4
HIV 97, 100–1, 104–11, 115
Hormone replacement therapy 71 *et passim*, 99

Informatics 5
  and health information 58 *et passim*
  Internet 5, 57, 62, 64, 67, 71–83, 98, 118, 161, 168–72
  and telemedicine 5, 84 *et passim*
Informed choice and consent 25, 33–35, 39
  and ambivalence 41–2, 49
Innovation dynamics 5, 85, 88, 113–14, 146 *et passim*
  and design 134–5, 147, 154–5, 160
  and drugs 186–93
  surgical 156–8
  systems 155–6
  and users 164, 165
Intraocular lens 146, 147–51
IVF 119, 212

Lay referral system 58, 64, 67

Medical Devices Agency 159, 207
Medicalisation 2, 114
Medicines Control Agency 207
MHRA 207

Narratives of health 2
National Institute for Clinical Excellence 120, 159, 233
New molecular entities 178, 184, 185
NHS Direct/Online 61–7, 93, 115, 135

Pain relief 141, 143–4, 236
Patient empowerment 5, 61, 85, 95, 97, 121, 125, 133
Patient as expert 95, 97, 98–100, 105–6, 111
Patient expertise 97 *et passim*
Patient health pathways 57 *et passim*
Pharmaceuticals 98, 120, 156, 157, 177–93
and regulation 180–3, 184
Population screening 11, 14
Pre-implantation genetic diagnosis 40, 42
Prenatal testing and screening 25 *et passim*
Pro-ana movement 100–4, 109–10
Professional expertise 43–4, 64, 83, 121
and organisational change 133
Psychiatric disease 236
Psychiatry 113, 124
Public Accounts Commission 159

Quality of life 234, 237, 238–9, 240–1

Rational drug design 188–9, 192
Regulation 7, 158–9
and hybridity 195–7, 216
and innovation 179–83, 196
Reproductive technologies 144
Risk 2, 5, 11 *et passim*
risk society 22, 132
and telemedicine 86, 89–92

Sick role 23, 58, 62, 65
Stem cells 195, 211 *et passim*
and cultural trading 217
and ethics 219–20
Social exclusion 172
Surveillance medicine 2, 12, 99

Technical codes 113–15, 122–5
Technology-dependent children 136–40
Temporality and the body 137–40
Tissue engineering 5, 7, 195, 202–8, 209, 211
Toxicology 182–3
Trust in expertise 68

UXIRA 198, 199, 200–1

Wheelchairs 161–8

Xenotransplantation 195, 197–202, 208, 210